# SpringerBriefs in Physics

SpringerBriefs in Physics are a series of slim high-quality publications encompassing the entire spectrum of physics. Manuscripts for SpringerBriefs in Physics will be evaluated by Springer and by members of the Editorial Board. Proposals and other communication should be sent to your Publishing Editors at Springer.

Featuring compact volumes of 50 to 125 pages (approximately 20,000–45,000 words), Briefs are shorter than a conventional book but longer than a journal article. Thus, Briefs serve as timely, concise tools for students, researchers, and professionals.

Typical texts for publication might include:

- A snapshot review of the current state of a hot or emerging field
- A concise introduction to core concepts that students must understand in order to make independent contributions
- An extended research report giving more details and discussion than is possible in a conventional journal article
- A manual describing underlying principles and best practices for an experimental technique
- An essay exploring new ideas within physics, related philosophical issues, or broader topics such as science and society

Briefs allow authors to present their ideas and readers to absorb them with minimal time investment.

Briefs will be published as part of Springer's eBook collection, with millions of users worldwide. In addition, they will be available, just like other books, for individual print and electronic purchase.

Briefs are characterized by fast, global electronic dissemination, straightforward publishing agreements, easy-to-use manuscript preparation and formatting guidelines, and expedited production schedules. We aim for publication 8–12 weeks after acceptance.

More information about this series at http://www.springer.com/series/8902

Mikhail I. Dyakonov

# Will We Ever Have a Quantum Computer?

 Springer

Mikhail I. Dyakonov
Laboratoire Charles Coulomb
Université de Montpellier
Montpellier, France

ISSN 2191-5423 ISSN 2191-5431 (electronic)
SpringerBriefs in Physics
ISBN 978-3-030-42018-5 ISBN 978-3-030-42019-2 (eBook)
https://doi.org/10.1007/978-3-030-42019-2

This Springer imprint is published by the registered company Springer Nature Switzerland AG
The registered company address is: Gewerbestrasse 11, 6330 Cham, Switzerland

# Preface

This book about quantum computing is mostly addressed to non-specialists: physicists and engineers, university and high school students, as well as to anybody who might be interested in understanding the current state of the art and the perspectives for this field. It could also be of some help to scientific journalists writing about quantum mechanics and quantum computing. As the reader might have guessed, my answer to the question in the title is negative, and the reasons for such a point of view are explained below, mostly in non-technical terms.[1]

During more than 20 years we have been witnessing an overwhelming worldwide rush towards "quantum computing" accompanied by an unprecedented level of hype. One can hardly find an issue of any science digest magazine, or even of a serious physical journal, that docs *not* address quantum computing. Funds are generously distributed, Quantum Information Centers are opening all over the globe, and very soon the happy Kingdom of Bhutan in the Himalayas will be the only country without such a Center.[2] Breathtaking perspectives are presented to the layman by enthusiastic scientists and even more enthusiastic journalists. The impression has been created that quantum computing is going to be the next technological revolution of the twenty-first century.

Many researchers feel obliged to justify whatever research they are doing by claiming that it has some relevance to quantum computing. Computer scientists are proving and publishing new theorems related to quantum computers at a rate of *one article per day*. A huge number of proposals have been published for various physical objects that could serve as quantum bits or *qubits*. The general public, as well as the funding agencies, have been convinced that the quantum revolution

---

[1]Some readers might say: But we already have quantum computers that are fabricated and sold by the D-wave company, so what's the point? To avoid misunderstanding, I stress from the start, that this book is about *universal gate-model quantum computing* to which the overwhelming majority of the existing literature is devoted. While some quite interesting results have been obtained with D-wave machines, they still do not qualify to be called "quantum computers" in the generally accepted sense of this term.

[2]Note that so far no useful quantum information whatsoever is available.

bringing quantum finance, quantum Internet, quantum security against terrorism, and other quantum miracles, is at our doorstep.

Here is just one of many very recent examples of this sort of promises, this time from IBM: "Hyper-accurate long-term weather forecasting. Life-saving drugs discovered through deep study of the behavior of complex molecules. New synthetic carbon-capturing materials to help reverse climate change caused by fossil fuels. Stable, long-lasting batteries to power electric vehicles and store green energy for the utility grid [1]."

The word "quantum" has entered households and rock stars' vocabulary and has acquired the meaning of wonderful, modern, powerful, cool, and awesome (The founders of quantum mechanics would have been very surprised by that).[3] Numerous conferences, workshops, and symposiums are held every year worldwide assembling thousands of researchers. The science of "Quantum information and quantum computation" is being taught in many universities and engineering schools around the world, attracting the best students who dream of contributing to this daring enterprise.

A revealing precedent for this state of collective mind can be found in the (incomparably smaller scale) medieval quest for the Holy Grail described by different authors as a sacred treasure in the form of a cup, dish, or stone with miraculous powers that provide happiness, eternal youth, or sustenance in infinite abundance—quite similar to what is expected from eventual quantum computing. As we now know, the Holy Grail was never found...

As of January 2020, Google search gives 8,050,000 results for "quantum computing" and 242,000 results for "quantum computing with" (proposals for various physical objects that can serve as *qubits*, the building blocks of eventual quantum computers), and these numbers increase every day.

It has become something of a self-perpetuating arms race, with many organizations and institutions seemingly staying in the race if only to avoid being left behind. Some of the world's top scientists, at Google, IBM, Intel, Microsoft, and others, are working hard, and with lavish resources in state-of-the-art laboratories, to realize their vision of a quantum computing future.

In light of this global frenzy, it is natural to wonder: When will useful quantum computers be constructed? The most optimistic experts estimate it will take about 10 years more. More cautious ones predict 20 to 30 years (Similar predictions have been voiced, by the way, during the last 20 years). My answer is: "Not in the foreseeable future" and this book is devoted to explaining such a point of view. Having spent decades doing research on quantum phenomena in atomic and condensed-matter physics, I have developed a very pessimistic view based on an understanding of the insurmountable difficulties that should be overcome to ever make quantum computing work.

---

[3]The legitimate use of the adjective "quantum" is to designate physical objects (which, as a rule, are of atomic or subatomic scale) and phenomena that obey the laws of quantum mechanics, as opposed to classical mechanics.

I belong to a tiny minority of the so-called "quantum computing skeptics" among physicists (some of them are Nobel prize laureates) and mathematicians, who have publicly expressed a similar view. Among them: Robert Alicki [2], Serge Haroche and Jean-Michel Raymond [3], Gerard 't Hooft [4], Subhash Kak [5], Gil Kalai [6], Rolf Landauer [7], Robert Laughlin [8], Leonid Levin [9], Stephen Wolfram [10], and a few others. However, their sober voices were well below the level of the enthusiastic noise created by the innumerable propagandists of quantum computing.

For the time being, there still does not exist a single quantum device capable of doing elementary arithmetic comparable to the capacities of an abacus, let alone surpassing the slide rule or the simplest electronic calculator, in any type of calculations.

I have tried to explain things as simple as possible (at a certain risk of over-simplification) to make the book accessible to readers with only a high school education in physics and mathematics. The reader can skip paragraphs that he finds difficult to understand, without losing the general picture.

Chapter 1 presents a brief history of the field and the basic ideas for quantum computing.

Chapter 2 presents the 2002 and 2016 roadmaps for quantum computing and a tiny part of the existing proposals for various physical objects that can serve as qubits.

Chapter 3 describes the basic quantum mechanics as applied to the simplest two-state quantum object, the electron spin.

Chapter 4 addresses the hypothetical quantum computer in the light of physical reality.

Chapter 5 considers some additional physical problems on the way to quantum computing.

Chapter 6 discusses the relation between physics and mathematics, some sociological issues, and my view on the prospects for quantum computing, summed up in conclusions.

Montpellier, France                                         Mikhail I. Dyakonov

# Contents

# Chapter 1
# Brief History and Overview of the Main Ideas for Quantum Computing

The idea of quantum computing was first put forward in 1980 by Benioff [11], who described a quantum mechanical model of a Turing Machine, and by the Russian mathematician Manin [12] who casually mentioned it in a rather vague form. In 1981, it was independently proposed by Feynman [13] (also in a quite vague form). Realizing that because of the exponential increase of the number of quantum states[1] computer simulations of quantum systems become impossible when the system is large enough, he advanced the idea that to make them efficient the computer itself should operate in the quantum mode: "Nature isn't classical, dammit, and if you want to make a simulation of Nature, you'd better make it quantum mechanical, and by golly it's a wonderful problem, because it doesn't look so easy".

In 1985, Deutsch [14] formally defined the universal quantum computer as a quantum analog of the universal Turing machine.

***Shor's algorithm.*** The subject did not attract much attention until 1994 when Shor [15] proposed an algorithm that (with an *ideal* quantum computer) could factor extremely large numbers much faster than the conventional classical computer. This outstanding mathematical result has triggered the whole field of quantum computing! The excitement was caused by the possibility for the eventual quantum computer to break security codes, some of which (but not all) are based on the enormous difficulty, or even impossibility, for conventional computers to factor 1000-digit numbers, which are products of very large primes.

***Grover's algorithm.*** Another famous quantum algorithm for database search was proposed in 1996 by Grover [16]. Here again, the ideal quantum computer should be much faster than a classical computer. Several other quantum algorithms have been also proposed.

***Qubits versus bits.*** Our *classical* computers are basically an assembly of very fast on/off switches, physically realized initially as vacuum tubes, and later - as tiny

---

[1] This meaning of this will be explained in Chap. 3

© The Author(s), under exclusive license to Springer Nature Switzerland AG 2020
M. I. Dyakonov, *Will We Ever Have a Quantum Computer?*,
SpringerBriefs in Physics, https://doi.org/10.1007/978-3-030-42019-2_1

transistors. The performance of the computer consists in operating these switches according to a prescribed program. Each switch represents a yes/no *bit* of information.

At a given moment the state of the *classical* computer is described by a sequence (10100110...), where 1 and 0 represent *bits* of information physically realized as the *on* and *off* states of individual transistors. With $N$ transistors, there are $2^N$ distinct possible states of the computer, which is quite a lot for large $N$. The computation process consists in a sequence of switching some transistors between their 1 and 0 states according to a prescribed program.

In *quantum* computing, the classical two-state element is replaced by a quantum element with two *basic* states, which is common to represent as $|1\rangle$ and $|0\rangle$[2] In the literature, an arbitrary quantum object with 2 basic states is called a "qubit" (= quantum bit).

The choice of the basic $|1\rangle$ and $|0\rangle$ states is a question of convention, and these states are *not the only ones* that are possible. The *general* state of the qubit is described by a *wave function*:

$$\psi = a|1\rangle + b|0\rangle, \tag{1.1}$$

where the *quantum amplitudes* $a$ and $b$ are arbitrary complex numbers satisfying the normalization condition

$$|a|^2 + |b|^2 = 1, \tag{1.2}$$

moreover, $|a|^2$ and $|b|^2$ give the *probabilities* for the qubit to be in the $|1\rangle$ and $|0\rangle$ states, respectively, see Chap. 3.

It is important to understand that the values of $a$ and $b$ depend on *our choice* of the basic states and that there is an infinite number of equivalent possibilities for doing this (similar to the infinite number of possibilities to choose the $x$ and $y$ axes in the plane).

In contrast to the classical bit that can be only in *one* of the two on/off states, (1) or (0), the qubit can be in a *continuum* of states defined by the quantum amplitudes $a$ and $b$. *The qubit is a continuous object!*

Two qubits have 4 basic states: $|11\rangle, |01\rangle, |10\rangle,$ and $|00\rangle$, thus the general state of 2 qubits is described by the wave function

$$\psi = a|11\rangle + b|01\rangle + c|10\rangle + d|00\rangle, \tag{1.3}$$

with 4 quantum amplitudes $a, b, c,$ and $d$, subject to the normalization condition $|a|^2 + |b|^2 + |c|^2 + |d|^2 = 1$, so that e.g. $|a|^2$ is the *probability* that measurement will give the result (11), and similarly for other states.

With $N$ qubits, there are $2^N$ *basic* states, thus the *general* state of $N$ qubits is defined by a wave function with $2^N$ quantum amplitudes, which are arbitrary complex numbers restricted by the normalization condition only.[3] In contrast, a *classical* system

---

[2] Another notation used for these states is: $|up\rangle$ and $|down\rangle$.

[3] This is just text-book quantum mechanics. Apparently, it comes as a surprise for many theorists of quantum computing.

**Fig. 1.1** The widespread mystical view of a qubit. The quantum switch described by the wave function $\psi$ is supposed to be "on" and "off" *at the same time*

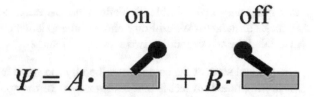

of $N$ pointers that can freely rotate around fixed points, is defined by $2N$ parameters (two polar angles for each pointer). For example, 10 pointers are described by 20 continuous parameters, while 10 qubits are characterized by $2^{10} = 1024$ continuous parameters.

Thus, by definition, the hypothetical quantum computer is an ***analog machine***, because it is supposed to operate with *continuous* parameters, in contrast to our conventional *digital* computers operating with *discrete* parameters (yes/no, or on/off).

***Up and down at the same time?*** The simple Eq. (1.1) is a source of profound misunderstanding, leading to the mystical and frightening statement, widely cited by practically every journalist commenting on quantum computing, that the qubit (spin) can exist **simultaneously** in *both* of its $|1\rangle$ and $|0\rangle$ states, see Fig. 1.1.

This is like saying that vector in the $xy$ plane directed at $45°$ to the $x$-axis *simultaneously* points *both* in the $x$- and $y$-directions. The journalists are not supposed to understand quantum mechanics, they are only repeating what they are told by experts, who *should* understand it. Since this "at the same time" is a widespread mystification, it is worth a short discussion in some detail.

Consider first a *classical* vector $R$ of unit length, which can freely rotate in the $xy$ plane around a fixed point with coordinates $x = 0$, $y = 0$. Let $n_x$ and $n_y$ be unit vectors pointing in the $x$- and $y$- directions respectively, so that

$$R = Xn_x + Yn_y. \tag{4}$$

Thus the projections of our vector on the $x$ and $y$ axes are: $X = \cos \varphi$, $Y = \sin \varphi$, where $\varphi$ is the angle that the vector $R$ makes with the $x$-axis.

Clearly, it is *our* choice which direction to label as the $x$-axis, and the coordinates $X$ and $Y$ of our pointer depend on this choice. For example, we could choose the direction of the $x$-axis along $R$, then we would have $X = 1$, $Y = 0$.

It would be quite bizarre to interpret Eq. (4) by saying that our vector $R$ points in the $x$ and $y$ directions *at the same time*! Rather, it generally has *projections* on both the $x$- and $y$-axes and these projections depend on *our* choice of these axes.

The same is true even in the "magical" world of quantum mechanics, where Eq. (1) is quite analogous to Eq. (4)—with an important difference concerning the possible results of *measurements*, see below.

***Electron spin as qubit*** The simplest quantum object with 2 basic states is the electron internal angular momentum, called *spin*, which has the peculiar quantum property of having only *two* possible projections *on any **axis***: $+1/2$ or $-1/2$ (in units of the Planck constant $\hbar$). We arbitrary chose some axis which we label as the $z$-axis, and also label the corresponding two states as $|1\rangle$ and $|0\rangle$.

The general state of the spin is described by the wave function in Eq. (1). The *truly surprising* quantum fact is that *measurement* of the spin projection on *any* chosen axis will give either $+1/2$ or $-1/2$ with certain probabilities, see Chap. 3. Before a measurement, there is no way to know or predict which of the two possibilities will be chosen, only the corresponding probabilities $|a|^2$ and $|b|^2$ are available.

It can be shown, that if $|1\rangle$ and $|0\rangle$ correspond to $+1/2$ or $-1/2$ projections on the $z$-axis, then in the particular case when $a = b = 1/\sqrt{2}$, the spin points in the direction of the $x$-axis (meaning that in this case there is a 100% probability of measuring the $+1/2$ projection on the $x$-axis), rather than "pointing along and opposite to the $z$-direction at the same time", which is *absolutely meaningless*, even in strange world of quantum mechanics.

We know in advance that the spin projection on *any* chosen axis can be only $+ 1/2$ or $-1/2$, but only by making a measurement we can know which of the two possibilities has been realized, quite similar to a tossed coin: one wouldn't say that it is head and tails *at the same time*.

*Quantum gates*  This term means certain standard operations with individual qubits, or with some subset of qubits (unitary transformations, in mathematical language). Quantum gates change the quantum amplitudes $a, b, c,...$ in a manner that *we* choose. For example, a one-qubit quantum gate can change the state of an individual qubit from $|1\rangle$ to $|0\rangle$ or vice versa. A two-qubit gate can transform the state $|10\rangle$ to $|01\rangle$, etc.. Obviously, like any operation with continuous parameters, this never can be done exactly, but only with some limited precision.

*The idea of quantum computing* is to store information in the values of $2^N$ complex amplitudes describing the wave function of $N$ qubits, and to process this information by applying a sequence of quantum gates, that change these amplitudes in a precise and controlled manner, as well as by performing intermediate measurements. The final result is supposed to be read by *measuring* the state of all our qubits (i.e. the projection of our spins on some chosen axis).

The value of $N$ needed to have a useful machine, i.e. one that is supposed to beat a classical computer in solving a certain restricted class of problems, is estimated to be $10^3$–$10^5$ or more. Note that even $2^{1000} \sim 10^{300}$ is much greater than the number of elementary particles in the Universe (!), which is estimated to be only about $10^{80}$.

*Error correction*  In contrast to the on/off switch which, as we know from everyday experience, is very reliable and stable, the "quantum bits" or *qubits* are extremely fragile. Being continuous objects, they will spontaneously change their state (i.e. the values of the quantum amplitudes) because of all sort of noise produced by external random fields, interaction with other qubits, and unavoidable uncertainties of any our actions (gates), and this will totally disorganize the functioning of the eventual quantum computer. Indeed, the $\sim 10^{300}$ or more continuously changing quantum amplitudes of the grand wave function describing the state of the computer must closely follow the desired evolution imposed by the quantum algorithm. The random drift of these amplitudes caused by noise, gate inaccuracies, unwanted interactions, etc., should be efficiently suppressed.

In response to those quite evident requirements, Shor [17] and Steane [18] proposed a method of *quantum error correction*, based on redundancy, which is a generalization of the error correction method in conventional (classical) computers, but much more sophisticated. It is widely recognized that **quantum computing without efficient error correction is impossible**.

Experimentally, there are several papers describing "proof of principle" error correction of a *single* qubit, the most recent, rather modest, result [19] consists in increasing the lifetime of single corrected qubit to 320 microseconds, which is about 2 times longer than the lifetime of an uncorrected qubit. So far, 25 years since the concept of quantum error correction was advanced, *there still does not exist any quantum device, which performs efficient error correction, even on a very small scale*.

**Threshold theorem.** It is not obvious at all that error correction can be done, even in principle, in an analog machine which state is described by at least $10^{300}$ continuous variables. Nevertheless, it is generally believed that the prescriptions for fault-tolerant quantum computation using the technique of error-correction by encoding [18, 20] and concatenation (recursive encoding) give a solution to this problem.

By active intervention, errors caused by noise and gate inaccuracies are supposed be detected and corrected during the computation. The so-called "threshold theorem" by Aharonov and Ben-Or [21] says that, once the error per qubit per gate is below a certain value estimated as $10^{-6}-10^{-4}$, indefinitely long quantum computation becomes feasible, at a cost of substantially increasing the number of qubits needed. However, very luckily, the number of qubits increases *only polynomially* with the size of computation, so that instead of 1000 qubits, we must use maybe a million (or possibly, a billion) qubits only.

Thus, the theorists claim that the problem of quantum error correction has been solved, at least in principle, so that now physicists and engineers need only to work hard on finding the good candidates for qubits and on approaching the accuracy required by the threshold theorem:

"As it turns out, it is possible to digitize quantum computations arbitrarily accurately, using relatively limited resources, by applying quantum error-correction strategies developed for this purpose" [22].

"The theory of fault-tolerant quantum computation establishes that a noisy quantum computer can simulate an ideal quantum computer accurately. In particular, the quantum accuracy threshold theorem asserts that an arbitrarily long quantum computation can be executed reliably, provided that the noise afflicting the computer's hardware is weaker than a certain critical value, the *accuracy threshold*" [23].

All the hopes for scalable quantum computing rely entirely on the threshold theorem. We will have a closer look at this theorem in Chap. 4.

***Universal quantum computer.*** This is a theoretical concept of a machine that combines the full power of a classical computer with the power of a quantum computer, and enables, among other things, simulation of physics, especially quantum mechanics, in the spirit of the initial ideas of Benioff, Feynman, and Deutsch. The universal quantum computer is supposed to use a universal quantum logic gate set and to beat classical computers by successfully performing Shor's factoring algorithm, Glover's search algorithm, etc. In spite of numerous optimistic announcements during the last 20 years, this ultimate goal permanently remains 20 or more years in the future (see Chaps. 4 and 6).

***Quantum annealing and beyond.*** Understanding the enormous difficulties of the conventional *gate model* of quantum computing, schematically described above, the Canadian company D-wave Systems founded in 1999 by Haig Farris and Geordie Rose proposed and realized a quite different idea of a quantum machine [24, 25]. This approach, now followed and developed by IBM, Google, Microsoft, and others, is based on using tiny superconducting rings with Josephson junctions [26] at ultralow dilution refrigerator temperatures in the millikelvin range.

The Josephson junction is formed by two superconductors separated by a thin potential barrier across which the electrons can tunnel (a well known quantum-mechanical phenomenon). Because of this tunneling, a superconducting ring with a Josephson junction can indefinitely support current [27].

The magnetic flux, i.e. magnetic field times the surface enclosed by the ring, is *quantized,* which means that it can be equal only to an integer number ($0, \pm 1, \pm 2 \ldots$) of elementary flux quanta $\Phi_s = \pi \hbar c/e$. Here plus or minus designate the direction of the electric current in the ring, $\hbar$ is the Planck constant, $c$ is the light velocity, $e$ is the electron charge. (There is about ten millions of elementary magnetic flux quanta per square centimeter in the very weak magnetic field of the Earth.).

Depending on some parameters of the system, Josephson junctions can operate either as classical two-state bits (and classical computing using Josephson logic have been demonstrated [28], however the need for liquid Helium temperatures and lower makes it impractical), or as quantum bits.

One can use junctions with e.g. 0 and 1 flux quanta (or better, $+1$ and $-1$, since—in the absence of external magnetic field—those states have equal energies[4]), as a qubit, and these states are very stable, but one can go from one state to another by applying appropriate microwave pulses. To make neighboring rings interact, one connects them by superconducting *couplers.*

*Annealing* is a term coming from metallurgy. After initial preparation, any system whether classical or quantum, when slowly cooled down, tends to its ground a state with the lowest possible energy. Within a D-wave machine, this ground state depends on the way the superconducting rings are interconnected, and one can measure *some* of its parameters. The interconnections should be such that this ground state represents a solution of some problem of interest. (An optimization problem is analogous

---

[4]The importance of having equal energies of the $|1\rangle$ and $|0\rangle$ states of a qubit is explained in Chap. 4.

to the search of a minimum in a landscape of peaks and valleys, the best solution corresponding to the lowest point in the landscape).

Annealing might provide some knowledge on the behavior of large and complicated quantum ensembles and also be useful for some special tasks in modeling, optimization, and sampling. A remarkable simulation of the Kosterlitz-Thouless phase transition was recently demonstrated with a D-wave machine in a network of Josephson superconducting rings arranged in a frustrated lattice [29].

While certainly being a *partly* controlled quantum system, such a machine, is very far from being a universal quantum computer: it only computes its own ground state, in the same way as a piece of melted iron, when slowly cooled down, "computes" its ground, or some meta-stable, state (which might be extremely difficult, or even impossible to calculate theoretically or by classical computer simulations).

*Beyond annealing.* During the last decade, Google, IBM, and others have purchased and somewhat modified and improved the D-wave machines, learned to partially control them, and have obtained some interesting and potentially useful results which however are still extremely far away from the anticipated sweeping quantum revolution.

To achieve the ultimate goal of building a universal quantum computer one must learn to implement reliable and very precise two-qubit quantum gates, which needs incorporation of many Josephson junctions in a network of macroscopic LC circuits, and much hard work has been done in this direction, see e.g. Ref. [30]. One also needs to implement efficient error correction, and this task is extremely far away.

Last year, Google has announced that they have reached "quantum supremacy" [31] in a rather artificial task, a claim that was immediately questioned by the rival IBM [32] and others.

Quite recently (January 2020), Google's large team of researchers headed by John Martinis has announced a really important breakthrough: the demonstration of a continuous set of two-qubit gates [33], thus somewhat approaching the ultimate goal of creating a true quantum computer.

Whatever may be the future perspectives of this approach, for the time being, it appears to be the *only one* that actually works (in the sense: giving at least, *some* interesting results that can be analyzed), although the research is still very far away from attaining continuous error correction, which is absolutely indispensible for achieving quantum computing. Not one of the thousands of other published proposals of the type "quantum computing with" (see Chap. 2) has ever been realized so far, and most probably, never will.

# Chapter 2
# Roadmaps, Panels, and Proposals: Quantum Computing with…

In Europe, a group of distinguished scientists has addressed the European Commission with a "Quantum manifesto" (May, 2016), a *call to launch an ambitious European initiative in quantum technologies, needed to ensure Europe's leading role in the second quantum revolution now unfolding worldwide* [34]. The Commission appeared to be quite sensitive to this call[1] and allocated 1 billion Euros for approaching this goal. The anticipated results for quantum computing are:

*0–5 years*: Operation of a logical qubit protected by error correction or topologically New algorithms for quantum computer

*5–10 years*: Small quantum processor executing technologically relevant algorithms. Solving chemistry and materials science problems with special purpose quantum computer

*>10 years*: Integration of quantum circuit and cryogenic classical control hardware. General purpose quantum computers exceed computational power of classical computers

This is not the first "quantum" roadmap. Previously, in 2002, at the request of the Advanced Research and Development Activity (ARDA) agency of the United States government, a team of distinguished experts in quantum information had already established a similarly audacious roadmap [35] for quantum computing with the following five- and ten-year goals:

by the year 2007, to

- encode a single qubit into the state of a logical qubit formed from several physical qubits
- perform repetitive error correction of the logical qubit, and

---

[1] The Americans are spending many billions of dollars on this, they wouldn't do anything entirely stupid, would they?

© The Author(s), under exclusive license to Springer Nature Switzerland AG 2020
M. I. Dyakonov, *Will We Ever Have a Quantum Computer?*,
SpringerBriefs in Physics, https://doi.org/10.1007/978-3-030-42019-2_2

- transfer the state of the logical qubit into the state of another set of physical qubits with high fidelity, and

by the year 2012, to implement a concatenated[2] quantum error-correcting code.

While a *very* benevolent jury could consider the first of the 2007 goals to be *partly* achieved by now, the expectations for the other 2007 goals, and especially for the 2012 goal, are *wildly off the mark*. So are some other predictions: "As larger-scale quantum computers are developed over the next five and ten years, quantum simulation is likely to continue to be the application for which quantum computers can give substantial improvements over classical computation".

However, nothing even remotely resembling larger-scale quantum computers, nor any improvements over classical computation, have occurred so far.

Apparently, going from 5 qubits to 50 (the goal set by the ARDA Experts Panel roadmap for the year 2012!) presents hardly surmountable experimental difficulties. Most probably they are related to the simple fact that $2^5 = 32$, while $2^{50} = 1125899906842624$ (see Chap. 1 for the explanation of exponential increase of the difficulty with increasing the number of qubits).

One can notice a similarity between the 5- and 10-year projections in the 2002 and the 2016 roadmaps, especially that the 2016 Quantum Manifesto five-year goal is in fact the same as the 2002 ARDA five-year goal. Will we see another similar roadmap in the year 2030, with a 5-year goal of encoding single qubits into the state of a logical qubit?

More recently, in late 2018, another expert panel assembled by the U.S. National Academies of Science, Engineering and Medicine issued a detailed 205-page report discussing some of the challenges facing quantum computing as a technology of practical value [36]. The authors of the report dryly state that within the next decade *no quantum computer will be capable* of breaking cryptographic codes based prime number factoring (an example of a task that quantum computers are supposed to be especially well suited for), and do not provide any opinion on whether or not this will be possible in a more distant future.

*Proposals*  The multitude of proposals for different ways to do quantum computing, as well as of various physical objects that can serve as qubits, is truly amazing. A simple list of such proposals with abstracts would require the entire space of this book. To show the reader just the tip of the iceberg, below is a small number of randomly picked proposals (source: *arxiv.org*, references can be easily found there). For more information the reader is invited to google "quantum computing with".

*Quantum computing with*:

- non-deterministic gates
- bosonic atoms
- highly verified logical cluster states
- Pfaffian qubits

---

[2]"Concatenation" in this context means spreading the information of one qubit onto a certain state of several qubits, so that the logical qubit gets encoded by several physical qubits [17, 18].

- hyperfine clock states
- four-dimensional photonic qubits
- quantum-dot cellular automata in dephasing-free subspace
- generalized binomial states
- 1D projector Hamiltonian
- quantum-dot spin qubits inside a cavity
- graphene nanoribbons
- alkaline earth atoms
- Jaynes-Cummings model
- doped silicon cavities
- Read-Rezayi states
- electron spin ensemble
- ultra narrow optical transition of ultra cold neutral atoms in an optical lattice
- $p$-wave superfluid vortices
- railroad-switch local Hamiltonians
- global entangling gates
- semiconductor double-dot molecules
- decoherence-free qubits
- superqubits
- defects
- devices whose contents are never read
- alkaline-earth-metal atoms
- ionic Wigner crystals
- nanowire double quantum dots
- waveguide-linked optical cavities
- orbital angular momentum of a single photon
- probabilistic two-qubit gates
- non-deterministic gates
- small space bounds
- interaction on demand
- perpetually coupled qubits
- only one mobile quasiparticle
- moving quantum dots generated by surface acoustic waves
- para-hydrogen
- programmable connections between gates
- incoherent resources and quantum jumps
- $\nu = 5/2$ fractional quantum Hall state
- spin ensemble coupled to a strip line cavity
- vibrationally excited molecules
- Kerr-nonlinear photonic crystals
- atoms in periodic potentials
- Heisenberg ABAB chain
- endohedral fullerenes
- harmonic oscillators.......

Isn't this wonderful? Apparently, there is nothing at all in the physical world that is NOT suitable for quantum computing! However there were no attempts to realize any of those, or thousands of other, proposals so far, and it doesn't look likely that this will happen anytime in the future. Making such proposals, and writing corresponding articles, is quite easy: just look around and propose whatever you see, or have heard of, as a new candidate for a qubit...

From the sociological point of view, it would be very interesting to understand the reasons for the publication of these and thousands of other similar proposals.

Let us consider just one of the proposals listed above, "quantum computation with the $\nu = 5/2$ fractional quantum Hall state" [37]. The experimentally observed $\nu = 5/2$ quantum Hall plateau is unlike all the others (for which the denominator of the filling factor $\nu$ is odd) and it does not fit into the conventionally adopted composite fermions concept. Some people think that this is a manifestation of *anyons*, hypothetical particles intermediate between bosons and fermions that may exist in two dimensions. (See also the *Anyon theory of high $T_c$ superconductivity* [38], now completely forgotten). Others think differently, and nobody really knows.

Hence the *obvious* proposal: to use these hypothetical anyons for quantum computing [39]. We must move them around one another producing complex topological structures, so that knot theory may be used. The great advantage is that topological structures are intrinsically protected against noise [40].

The disadvantages are mostly on the practical side. The quantum Hall plateaus are observed in high magnetic field with a roughly $3 \times 6$ mm two-dimensional sample immersed in liquid Helium. The sample has several wires attached, typically 6, to apply and measure voltages and currents. The number of electrons in the sample is about $10^{10}$, and presumably it contains a similar number of anyons. So how are we supposed to create topological structures with these hypothetical anyons, just by applying voltages to the 6 (or, if one insists, even 6000) wires at our disposal?

However, this is an old story. Later, it was proposed that the hypothetical Majorana fermions, rather than the hypothetical anyons, are responsible for the 5/2 Hall plateau [41]. Consequently, it became obvious that quantum computing with Majorana fermions is extremely promising [42].

A by-product of this frenetic activity is that every physical object has become a qubit, independently of whether it is regarded in the quantum computing context, or not:

*Electron spin qubit*
*Hole spin qubit*
*Nuclear spin qubit*
*Josephson superconducting qubit*
*Cavity photon qubit*
*Trapped ion qubit*
*Para-Hydrogen qubit*
*Heisenberg ABAB chain qubit*
*etc.*

This looks pretty, like some modern poetry. So, instead of saying like in the good old days, "We are studying nuclear spin resonance", now one should say: "We are studying decoherence of nuclear spin qubits", thus implying that our study is directly related to the big problems of the day.

# Chapter 3
# Basics of Quantum Mechanics

To understand what quantum computing is about, one must necessarily have some idea about quantum mechanics. A typical textbook on this subject contains 300–400 pages and requires previous knowledge of classical theoretical mechanics and electrodynamics, as well of rather advanced mathematics. Here, the main ideas of quantum mechanics will be presented in a somewhat simplified and sketchy form, hopefully understandable to some extent by more general readers.

*The Planck constant* The experimental findings at the end of the 19-th century revealed some phenomena that could not be explained within the existent understanding of physics (now referred to as *classical* physics). One of them was related to the mystery of thermal radiation by arbitrary bodies at a given temperature $T$.

The existing (now called *classical*) electrodynamics and thermodynamics predicted an *infinite* radiation intensity arising from the high frequency part of the spectrum, and this paradox became the starting point for the new physics. To deal with this problem, Planck advanced the revolutionary hypothesis (1900) that electromagnetic radiation is emitted by portions, *quanta*, the energy $E$ of a single quantum being proportional to $\omega$, the radiation frequency: $E = \hbar\omega$, where $\hbar$ is the Planck constant.[1]

Since the average energy $E$ is proportional to the temperature $T$, the typical frequency of emitted radiation is $\omega \sim T/\hbar$ (here temperature is measured in units of energy), and emission with frequencies such that $\hbar\omega > T$ is strongly suppressed. Thus, because of quantum mechanics, there is a high frequency cutoff in the emitted radiation. (Previous application of classical mechanics and electrodynamics resulted in the so-called "ultra-violet catastrophe"—an infinite emission of energy in the short wavelength, i.e. high frequency, part of the spectrum).

---

[1]It is habitual and more convenient to use $\hbar = h/(2\pi)$, where $h$ is the constant originally introduced by Planck.

© The Author(s), under exclusive license to Springer Nature Switzerland AG 2020
M. I. Dyakonov, *Will We Ever Have a Quantum Computer?*,
SpringerBriefs in Physics, https://doi.org/10.1007/978-3-030-42019-2_3

Starting with Planck's hypothesis, which not only was experimentally found to be true, but has explained a huge number of other physical phenomena, in the 1920s a theory was developed, *Quantum mechanics* (replacing the Newton classical mechanics), and later also *Quantum Electrodynamics* (replacing Maxwell's classical electrodynamics), that turned out to be extremely successful in explaining the structure and properties of atoms, molecules, nuclei, and condensed matter. In fact, "replacing" is not a quite correct term: classical physics obviously continues to be important and quite sufficient in many domains where passing to the limit $\hbar \to 0$ does not give diverging results (such a divergence would mean that quantum mechanics is indispensable, like in the problem of thermal radiation).

In particular, it explains the very existence of atoms. Consider the simplest Hydrogen atom, where a negatively charged electron rotates around the positively charged proton, similar to a planet rotating around the Sun. Classical physics says that an oscillating, or rotating, electric charge should emit radiation, thus losing its energy and eventually falling on the proton (like a satellite eventually falls on the Earth—for a different, but similar reason: loosing energy by friction in the atmosphere). Classical calculations show that such a collapse of atoms—and hence, of our entire world—would occur in about 1 ns ($10^{-9}$ s) after the creation of the Universe!

Quantum mechanics explains why this does not happen: the electron in a Hydrogen atom has a certain *ground state* with minimal energy[2] $E_0 = -me^4/(2\hbar^2)$, where $m$ and $e$ are the electron mass and charge respectively. Emission of radiation, i.e. losing energy, in this state is not possible—simply because no states with lower energy exist!

**The classical limit**   Note, that *quantum-mechanical formulas necessarily contain the Planck constant* $\hbar$. Taking the limit $\hbar \to 0$ in any quantum formula, we obtain the classical result, and all quantum effects disappear in this limit.

Thus, quantum mechanics is superior to classical mechanics in the sense that it contains classical mechanics as a limiting case ($\hbar \to 0$) where quantum effects become small and unimportant. (Similarly, Einstein's theory of relativity contains Newton mechanics as a limiting case $c \to \infty$, when the light velocity $c$ can be considered as being infinitely large with respect to velocities of interest).

In the above formula for $E_0$ this classical limit ($\hbar \to 0$) is: $E_0 = -\infty$, which corresponds to the electron *falling* on the proton, as the classical theory predicts, and this is a clear indication that quantum mechanics is indispensible for atomic physics.

Evidently, there are plenty of purely *classical* phenomena in which the elementary quantum is of no importance, e.g. the frequency $\omega$ of rotation of the wheels of your car is so low that the energy quantum $\hbar\omega$ is absolutely negligible, compared to the total mechanical energy of the rotating wheel.

Among other consequences, this means that there should be a condition for the possibility of the *quantum* computer, which evidently cannot exist in a purely classical

---

[2]The minus sign in this formula corresponds to the choice of $E = 0$ for the situation where both the electron and the proton are at rest and infinitely separated in space. The energy of the pair becomes *lower* (i.e. negative) when they form a Hydrogen atom.

world, i.e. in the limit $\hbar \to 0$. Thus, such a condition should have the form of $X < \hbar$. The expression and the physical meaning for this $X$, that should have the dimension of energy $\times$ time (like the Planck constant $\hbar$), is presently *unknown*, which appears rather awkward...

There is also a related interesting question: What is the classical limit (at $\hbar \to 0$) of the hypothetical quantum computer? One might think that this limit is provided by our good old classical digital computer. This is not correct! In fact, the classical limit of quantum computer would be a classical *analog* computer (see Chap. 4).

**Measurements** In quantum mechanics, the result of a measurement generally cannot be exactly predicted *in principle*, only the probabilities of specific results can be known beforehand.[3] Those probabilities are given by the squared modules of quantum amplitudes. Thus, in a state described by Eq. 1 the quantities $|a|^2$ and $|b|^2$ give the probabilities for measuring the qubit in $|1\rangle$ and $|0\rangle$ states, respectively. There is *no way* to predict the result of an individual measurement, except for the special case when $|a|^2 = 1$, and $|b|^2 = 0$ (or vice versa).

**Spin** It happens that the main constituents of matter: electron, proton, neutron have an internal angular momentum, *spin*. In an analogy to a rotating tennis ball, one can intuitively visualize this fact as a result of a permanent rotation of the particle around its axis (arbitrary oriented). It is conventional to denote this internal angular momentum as $M = \hbar S$, where $S$ is the dimensionless vector (more precisely, it is a *pseudo*-vector), called *spin*.[4] Spin, the simplest of the known quantum objects, is one of the most popular candidates for qubits.

*The projection of electron spin on **any** axis can be either* $+1/2$ *or* $-1/2$ *only!* We can chose an arbitrary axis (which can be labeled as the $z$-axis), and denote these two states as $|1\rangle$ and $|0\rangle$. The general state of the spin is described by a *wave function*: $\psi = a|1\rangle + b|0\rangle$, where $a$ and $b$ are complex numbers (the quantum *amplitudes*) satisfying the normalization condition: $|a|^2 + |b|^2 = 1$, with $|a|^2$ and $|b|^2$ being the *probabilities* for measuring the spin in the $|1\rangle$ and $|0\rangle$ states, respectively. A system of $N$ spins has an infinite number (continuum) of states described by $2^N$ complex parameters, the quantum amplitudes.

This applies to *any* quantum objects with 2 basic states. In contrast, a system of *classical* two-state objects (like on/off switches) has $2^N$ well defined *discrete* states. A system of $N$ *classical* pointers, which can rotate around fixed axes, has an *infinite* number of possible states defined by $2N$ (not $2^N$) *continuous* parameters: 2 polar angles for each pointer. It is important to understand and keep in mind these differences.

---

[3]This was a subject of a lively discussion in the early days of Quantum Mechanics. The absolutely novel idea that the notion of "probability" applies not only to *ensembles* of measurements, but also to an *individual* measurement was accepted with great difficulty, expressed by the famous Einstein's skeptical sentence "God does not play dice". Nevertheless, apparently, this is exactly what happens.

[4]Note that the Planck constant, like the angular momentum $M$, has the dimension of coordinate $\times$ momentum.

We reiterate, that $a$ and $b$ depend on *our* choice of the $z$-axis and the basic states. We can always choose an axis which is parallel to the current direction of spin. In this case we will have $a = 1$, and $b = 0$.

Since, $a$ and $b$ are complex numbers with real and imaginary parts, restricted by the normalization condition, and since the overall phase of the wave function is irrelevant, we are left with $2 \times 2 - 2 = 2$ *real* parameters defining the spin state— exactly like in classical mechanics, where the direction of a vector is defined by 2 polar angles.

A system of 2 spins has $2^2 = 4$ basic states: $|11\rangle$, $|10\rangle$, $|01\rangle$, and $|00\rangle$. Accordingly, they are described by the wave function $\psi = a|11\rangle + b|10\rangle + c|01\rangle + d|00\rangle$ with 4 complex amplitudes $a, b, c, d$, restricted only by the normalization condition $|a|^2 + |b|^2 + |c|^2 + |d|^2 = 1$. Now, the number of independent *real* parameters describing this state is $2 \times 4 - 2 = 6$, while a classical description of two vectors with fixed length needs only $2 \times 2 = 4$ polar angles. The difference is a quantum effect due to *correlations* between the two spins.

*In the general case of N spins, the whole system has a **continuum** of states, defined by the values of $2^N$ complex amplitudes which are continuous variables restricted by the normalization condition only.*

***Schrödinger equation*** Established in 1926, the Schrödinger equation is the basis of Quantum Mechanics, playing the same role as Newton's equation in classical mechanics. Like in classical mechanics, the notions of energy $E$ and of the Hamiltonian $H$ are of primordial importance [43].

This equation allows:

(a) To find the possible stationary states of an arbitrary quantum system and the possible values of the total energy $E$ of the system (its *energy spectrum*, which might be discrete or continuous),

(b) To predict the evolution of a quantum system in time for given initial conditions.

All our (quite satisfactory!) understanding of quantum systems (atoms, molecules, solids, etc.) is based on the Schrödinger equation. This equation is non-relativistic, i.e. it deals with situations when the velocities of all particles are small compared to the velocity of light. The relativistic quantum equation describing electrons with arbitrary velocities was established by Paul Dirac in 1930.

For a system consisting of $N$ objects with 2 basic states (qubits, or spins 1/2) the Schrödinger equation results in a system of $2^N$ linear differential equations for the wave function $\psi$ of the general form:

$$i\hbar d\Psi(\alpha_1, \alpha_2, \ldots, \alpha_N)/dt = \Sigma H(\alpha_1, \alpha_2, \ldots, \alpha_N; \beta_1, \beta_2, \ldots, \beta_N; t)\psi(\beta_1, \beta_2, \ldots, \beta_N),$$
$$(\beta) \tag{3.1}$$

where the values of all $\alpha$ 's and $\beta$ 's can be only $+1/2$ or $-1/2$ (or simply + or –), thus there are $2^N$ discrete (but time-dependent) values of the wave function $\psi$ ($\alpha_1$, $\alpha_2 \ldots \alpha_N$), while the *Hamiltonian* $H$ is a $2^N \times 2^N$ time-dependent matrix. The sum over ($\beta$) in Eq. (3.1) symbolizes summation over the two values of each $\beta_1, \beta_2, \beta_3...$ Solving this system (which might be quite difficult, or even impossible, for large $N$) is **the only way to understand the time evolution of our system** under the influence of our actions (application of "quantum gates") and external regular, or random fields (noise).

Strangely enough, with extremely rare exceptions, the theorists of *quantum* computing appear to have absolutely no need of the basic tools and attributes of *quantum* mechanics, such as Schrödinger equation, wave function, energy spectrum (which defines the dynamics of any quantum system) and never use them.[5]

*Dynamics of one- and two-spin systems* We will now have a look at what the Schrödinger equation predicts for the simplest systems of one and two spins.

*One spin* has two basic states, $|1\rangle$ and $|0\rangle$. In the absence of external fields, these two states can refer to an arbitrary direction in space, and both spin states, as well as their superposition $\psi = a|1\rangle + b|0\rangle$ with arbitrary amplitudes $a$ and $b$ satisfying the normalization condition, are *stationary* states (they do not change in time).

However, in the presence of magnetic field $\mathbf{B}$, it is wise to define the states $|1\rangle$ and $|0\rangle$ with respect to the direction of this field. Then, since the electron spin is associated with a magnetic moment $\mu$, the *energies* of the two states will be different and can be chosen as $+E_0$ and $-E_0$, (where $E_0 = \mu B$), so that the amplitudes $a$ and $b$ become time-dependent[6]:

$$a(t) = a_0\exp(-i E_0 t/\hbar), b(t) = b_0\exp(+i E_0 t/\hbar).$$

This is the quantum-mechanical way for expressing the simple fact that, because of the existence of a magnetic moment, $\mu$, associated with spin, the spin makes a precession around the direction of magnetic field with frequency $2E_0/\hbar$. For electron spins in the (quite weak) magnetic field of the Earth this frequency is about 1 MHz! Should the hypothetical quantum computer, using electron spins as qubits, be shielded from the Earth's magnetic field, and if yes, with what precision? There are no answers to this obvious question.

*Two spins* have 4 basic states, and the system is described by the wave function $\psi = a|11\rangle + b|10\rangle + c|01\rangle + d|00\rangle$. with 4 complex amplitudes $a, b, c$, and $d$. Now the dynamics of the system, described by the time dependence of these amplitudes, becomes more complex, with several different frequencies of oscillations, especially if we take into account the unavoidable *interaction* between our spins.

This gives us an idea of what will happen if our system consists of 1000 (interacting) spins in an external magnetic field: the frequency spectrum will become extremely rich, and the system dynamics will become chaotic, much more complex than the dynamics of a boiling soup!

---

[5]In my opinion, this reveals an insufficient understanding of Quantum Mechanics.

[6]We recall that $\exp(ix) = \cos x + i\sin x$, where $i$ is the imaginary unit.

Meanwhile, the theorists of quantum computing seem to innocently believe that the system of qubits evolves exclusively when and if the *Future Quantum Engineer*[7] decides to apply some external actions ("gates") to his qubits.

***Spin relaxation, or decoherence***  While the relaxation of two-level quantum systems was thoroughly studied during a large part of the 20th century, and is quite well understood, in the quantum computing literature there is a strong tendency of mystifying the relaxation process and make it look as an obscure quantum phenomenon [44, 45]: "The qubit (spin) gets entangled with the environment..." or "The environment is constantly trying to look at the state of a qubit, a process called decoherence", etc.

In a way, this sophisticated description may be true, however it is normally quite sufficient to understand spin relaxation as a result of the action of fluctuating external fields described by a time-dependent Hamiltonian $H(t) = A(t)\sigma$, where $A(t)$ is a random vector function, and $\sigma_x, \sigma_y, \sigma_z$ are Pauli matrices. In simple words, the spin continuously performs precession around fluctuating in time magnetic fields.

In many cases, these are not real, but rather effective magnetic fields induced by some interactions. A randomly fluctuating field is characterized by its correlation time, $\tau_c$, and by the average angle of spin precession, $\alpha$, during time $\tau_c$. For the most frequent case when $\alpha \ll 1$, the spin vector experiences a slow angular diffusion. The root mean square (RMS) angle after time $t \gg \tau_c$ is $\varepsilon \sim \alpha(t/\tau_c)^{1/2}$. Hence the relaxation time is $\tau \sim \tau_c/\alpha^2$. If one chooses a time step $t_0$, such that $\tau_c \ll t_0 \ll \tau_c/\alpha^2$, it can be safely assumed that $\varepsilon \ll 1$, and that rotations during successive time steps are not correlated.

These random rotations persist *continuously* for all the qubits inside the quantum computer. It is important to understand that the wave function describing an arbitrary state of $N$ qubits will deteriorate much faster than each individual qubit does. The reason is that this wave function $\psi = A_0|000...00\rangle + A_1|000...01\rangle + ... + A_2^N |111...11\rangle$ describes complicated correlations between the $N$ qubits, and correlations always decay more rapidly. For simplicity, suppose that all qubits are subject to random and uncorrelated rotations around the $z$ axis only. Then the state of the $j$-th qubit will change during one step accordingly to the rule: $|0\rangle \rightarrow |0\rangle, |1\rangle \rightarrow \exp(i\varphi_j)|1\rangle$, where $\varphi_j$ is the random rotation angle for this qubit.

Because of this, the amplitudes $A$ will acquire phases $\Sigma\varphi_j$, where the sum goes over all qubits that are in the state $|1\rangle$ in a given term of $\psi$. The typical RMS value of this phase after one step is $\sim\varepsilon N^{1/2}$, thus the time it takes for unwanted phase differences between the amplitudes $A$ to become large, is $\sim N$ times shorter than the relaxation time for an individual qubit. For this reason, it seems that if we choose the time step so that $\varepsilon^2 = 10^{-6}$ (the most optimistic of existing estimates for the noise threshold), but the quantum state contains $10^6$ qubits, then the computer is likely to crash during one step. I am unaware of anybody discussing this problem.

---

[7]The *Future Quantum Engineer* is a mythical figure from the quantum computing epos, that is supposed to implement the existing theoretical ideas in some distant and undefined future.

# Chapter 4
# The Theoretical Quantum Computer and the Physical Reality

The formal definition of a quantum computer was put forward by Deutsch [14] 35 years ago:

A quantum computer is a set of $N$ qubits in which the following operations are experimentally feasible:

1.  Each qubit can be prepared in some known state $|0\rangle$.
2.  Each qubit can be measured in the basis $\{|1\rangle$ and $|0\rangle\}$.
3.  A universal quantum gate (or set of gates) can be applied at will to any fixed-size subset of qubits.
4.  The qubits do not evolve other than via the above transformations.

The "universal quantum gate" is reminiscent of an old Russian joke. The story goes that during World War II an inventor appeared with an idea of extreme military value. Since at the time quantum cryptography did not yet exist, the inventor insisted that they take him to the very top, that is, to Stalin.

–   So, tell me what it is about.
–   It's simple, comrade Stalin. You will have three buttons on your desk, a green one, a blue one, and a white one (see Fig. 4.1). If you press the green button, all the enemy ground forces will be destroyed. If you push the blue button, the enemy navy will be destroyed. If you push the white button, the enemy air force will be destroyed.
–   OK, sounds nice, but how will it work?
–   Well, it's up to your engineers to figure it out! I just give you the idea.

Quantum-mechanically, anything that may happen in this world is a unitary transformation. Thus the wave function of the destroyed ground force is related to that of intact ground force by a unitary transformation, and the same is true for the navy and the air force. Similarly, it is for the quantum engineers to figure out the practical implementation of the universal quantum gate.

© The Author(s), under exclusive license to Springer Nature Switzerland AG 2020
M. I. Dyakonov, *Will We Ever Have a Quantum Computer?*,
SpringerBriefs in Physics, https://doi.org/10.1007/978-3-030-42019-2_4

**Fig. 4.1** A schematic view
of Supreme Commander's
work desk after the
installation of the universal
military transformation
device

Consider the difficulty of constructing a *classical* equivalent of the universal quantum gate, which must be much, much easier. Is there any hope of ever learning to physically transform with high precision an arbitrary state of 1000 classical pointers, which can freely rotate around fixed points, into another prescribed state?

It is a great illusion to think that the task might be easier in quantum mechanics with 1000 quantum spins. In fact, it is much, much more difficult, because it requires controlling not $2 \times 1000$ (which is already a lot), but $2^{1000}$ continuous parameters!

***Classical limit of a quantum computer*** When considering quantum mechanical phenomena, it is always instructive to understand what happens in the *classical limit* ($\hbar \to 0$). The classical limit of the quantum computer is NOT a digital classical computer (as many people believe), but rather, a classical analog machine, that stores information in the mutual orientations of classical pointers (or classical magnetic moments, like compass needles).

The relation can be depicted as follows:

digital computer  $\to$  classical analog computer  $\to$  quantum (analog) computer

*The classical analog computer is the intermediate link between a quantum and a digital computer.* A slide rule (known also as a "slipstick") is a primitive classical analog computer with just *one* controllable parameter.[1]

---

[1]Invented by the English mathematician and clergyman William Oughtred in about 1662 (see: https://en.wikipedia.org/wiki/William_Oughtred), during 3 centuries the slide rule has reliably served to generations of scientists and engineers, before being replaced, only about 50 years ago, by electronic calculators and later—by modern computers.

Evidently, the practical possibility and utility of building a classical analog computer with, say, 1000 controllable parameters, are questionable, to say the least. However it is much, much easier than to build a true quantum computer with 1000 qubits: while our hypothetical classical analog machine will be described by 1000 continuous parameters, the state of a truly (but even more hypothetical) quantum computer will be defined by $2^{1000}$ continuous parameters. One can say that it is *exponentially more difficult* to build a quantum computer than a classical analog computer.[2]

***Theoretical and practical quantum computers*** We should make a clear distinction between the quantum computer of the abstract quantum information theory, based on the Deutsch's above definition, and the working quantum computer, between a thought experiment and a real experiment. For a mathematician, rotating a vector in a huge Hilbert space is a routine algebraic operation, while doing it in practice may be possible or not, depending on the required precision, the dimension of the space, and other conditions imposed by the real physical world.

The practical performance of a quantum computer is based on manipulating on a microscopic level and with an enormous precision a many-particle physical system with continuous degrees of freedom. Obviously, for large enough systems, either quantum or classical, this becomes impossible, which is why such systems belong to the domain of statistical, not microscopic physics. The question is whether a system of $N = 10^3 - 10^6$ qubits (spins), needed to beat the classical computer in solving a limited special class of problems, is large enough in this sense. Can we ever learn to control the $2^N$ amplitudes defining the quantum state of such a system?

Generally, there are many interesting and useful things one could accomplish with ideal machinery, not necessarily quantum. For example, one could become younger by exactly reversing all the velocities of atoms in our body (and in some near environment), or write down the full text of all the books in the world in the exact position of a single particle, or store information in the $10^{23}$ vibrational amplitudes of a cubic centimeter of a solid. Unfortunately, unwanted but inevitable, noise, fluctuations, and inaccuracies of our manipulations impose severe limits to such ambitions.

***Quantum error correction*** As mentioned in Chap. 1, Shor [17] and Steane [18] proposed the idea of quantum error correction—an ingenious method designed primarily for bypassing the so called "no cloning" theorem: an unknown quantum state cannot be copied. (At first glance, this theorem prevents us from checking whether there are errors in the quantum amplitudes, so that one can correct them).

However, this proposed method assumed that the additional (*ancilla*) qubits, the measurements, and the unitary transformations to be applied, remain ideal. It is said

---

[2]In the popular press, the term "exponential" has acquired the meaning of "extraordinary", "enormous", etc. In fact, it serves simply to define a function $y(x)$, such that $y \sim a^x$. The difficulty of building a quantum computer increases proportional to the number of continuous parameters that should be under control, which is $2^N$, where $N$ is the number of qubits. Indeed, this is an exponential dependence on $N$. In contrast, a classical analog computer based on $N$ pointers, needs only $2N$ parameters to be under control.

that this type of error correction is not fault-tolerant, whatever this may mean. (If the ancilla qubits are flawless, why not use them in the first place?) The ultimate solution, the *fault-tolerant quantum computation*, was advanced by Shor [44] and further developed by other mathematicians, see [45–49] and references therein.

Now, nothing is ideal: all the qubits are subject to noise, measurements may contain errors, and our quantum gates are not perfect. Nevertheless, the threshold theorem says that arbitrarily long quantum computations are possible, so long as the errors are not correlated in space and time and the noise level remains below a certain threshold. In particular, with error correction a single qubit may be stored in memory, i.e. it can be maintained arbitrarily close to its initial state during an indefinitely long time. (See Chap. 6 for a discussion of this issue).

This contradicts all our experience in physics. Imagine a pointer, which can freely rotate in a plane around a fixed axis. Fluctuating external fields cause random rotations of the pointer, so that after a certain relaxation time the initial position gets completely forgotten. (For electron spins the corresponding relaxation time is typically on the scale of nanoseconds).

How is it possible that by using only other identical pointers (also subject to random rotations) and some external fields (which cannot be controlled perfectly), it might be possible to maintain indefinitely a given pointer close to its initial position? The answer we get from experts in the field, is that it can work because of quantum mechanics: "We fight entanglement with entanglement" or, in the words of the Quantum Error Correction Sonnet by Gottesman [46],

> With group and eigenstate, we've learned to fix
> Your quantum errors with our quantum tricks.

This does look suspicious, because in the physics that we know, quantum-mechanical effects are more easily killed by noise than classical ones. Below is a slight divertissement relevant to our subject.

***Capturing a lion in a desert*** The scientific folklore knows a joke about specialists in various fields proposing their respective methods of capturing a lion in a desert. (For example, the Philosopher says that a captured lion should be defined as one who is separated from us by steel bars. So, let's go into a cage, and the lion will be captured). Here, we are concerned with the Mathematician's method:[3]

*The desert D being a separable topological space, it contains a countable subset S that is everywhere dense therein. (For example, the set of points with rational coordinates is eligible as S.) Therefore, letting $x \in D$ be the point at which the Lion is located, we can find a sequence $\{x_n\} \subset S$ with $\lim_{n \to \infty} \{x_n\} = x$. This done, we approach the point x along the sequence $\{x_n\}$ and capture the lion.*

This method, depicted in Fig. 4.2, assumes that the only relevant property of the Lion is to be located at a given point in 2D space. Note also that neither *time*, nor what can happen to the Lion and the Hunter during the process, is a point of concern.

---

[3]I thank Konstantin Dyakonov for providing the text below, and Ekaterina Diakonova for drawing the picture.

**Fig. 4.2** The mathematician's method for capturing a Lion in a desert

$$\forall\, x \in D,\ \exists\, \{x_n\}_{n=1}^{\infty} \subset S \mid \lim_{n \to \infty} x_n = x$$

And finally, it is not specified how the sequence $\{x_n\}$ should be chosen, nor what the limit as $n \to \infty$ could mean in practice. These points are left to be elaborated by the practical workers on the ground.

Certainly, mathematics is a wonderful thing, both in itself and as a powerful tool in science and engineering. However, we must be very careful and reluctant in accepting theorems, and especially technical instructions, provided by mathematicians in domains outside pure mathematics. Whenever there is a complicated issue, whether in many-particle physics, climatology, or economics, one can be almost certain that no theorem will be applicable and/or relevant, because the explicit or implicit underlying assumptions will never hold in reality. The Hunter must first explain to the Mathematician what a lion looks like.

***The threshold theorem*** It is universally recognized that it is impossible to build a quantum computer without implementing efficient error correction. The idea of error correction is based on a distinction between *logical* and *physical* qubits, each logical qubit is supposed to be *encoded* by a certain number of physical qubits. The necessary number of physical qubits per logical qubit is estimated to be around $10^3 - 10^5$, so that the total number of physical qubits needed for building a useful quantum computer should increase from $\sim 10^3$ to $\sim 10^6 - 10^8$ only.

The threshold theorem is in fact a technical instruction for fault-tolerant quantum computation, interspersed with lemmas proving that every step will work as it should. It tells us how, by using imperfect instruments, to control on a microscopic level a million of quantum objects, like spins, subject to various sources of relaxation and unwanted interactions.

The instruction is overwhelmingly complicated. Below is a short quotation from the famous 63-page threshold-theorem article by Dorit Aharonov and Michael Ben-Or [21] describing a small part of the proposed verification procedure for the so-called "cat state": $2^{-1/2}(|0000000\rangle + |1111111\rangle)$. This procedure itself is only a minor detail in the whole scheme providing the anticipated fault-tolerance.

> We now want to compute whether the bits in the cat state are all equal. For this, note that checking whether two bits are equal, and writing the result on an extra blank qubit, can be easily done by a small circuit which uses Toffoli and NOT gates. We denote this circuit by $S$, and also denote the qubits in the cat state by 1, …$l$. We add $l - 1$ extra blank qubits, and apply the circuit $S$ first from each even pair of qubits (e.g. the pair of qubits (1, 2), (3, 4)…), to one of the blank qubits; Then apply $S$ from each odd pair of qubits (e.g. the pairs (2, 3), (4, 5)…) to one of the remaining blank qubits. We get $l - 1$ qubits which are all 1, if no error occurred, indicating that all the qubits are equal. We then apply a classical circuit on all these qubits, which checks whether they are all 1, and write the result on an extra blank qubit, which is our check bit, and indicates that all the bits in the cat state are equal. We now want to use the cat state, but condition all the operations on the fact that the check bit is indeed 1. However, if we do this, an error in the check bit, might propagate to all qubits in the state we are trying to correct. Hence, to keep the procedure fault tolerant, we construct $m$ different check bits, one for each qubit in the state we are correcting. This is done using $m(l - 1)$ blank qubits, and applying all the operations above, where each operation is repeated $m$ times to $m$ different target qubits. Thus, we can verify that the cat state is of the form $C_0|0\rangle + C_1|1\rangle$, fault tolerantly. We can now condition all the operations done in the syndrome measurement involving the $i$-th qubit, on the $i$-th check bit.

This gives the reader an idea of what the theoretical quantum computing literature looks like.

Thus, the theorists claim that the problem of quantum error correction is resolved, at least in principle, so that physicists and engineers have only to do more hard work in finding the good candidates for qubits and approaching the accuracy required by the threshold theorem:

> As it turns out, it is possible to digitize quantum computations arbitrarily accurately, using relatively limited resources, by applying quantum error-correction strategies developed for this purpose. [22]

> The theory of fault-tolerant quantum computation establishes that a noisy quantum computer can simulate an ideal quantum computer accurately. In particular, the quantum accuracy threshold theorem asserts that an arbitrarily long quantum computation can be executed reliably, provided that the noise afflicting the computer's hardware is weaker than a certain critical value, the *accuracy threshold*. [23]

No mention of any restrictions.

Note, however, that the proposed procedure assumes that the state of the qubits can change only as a result of *our* actions. This is absolutely wrong! See Chap. 6 for a discussion of this point.

***What precision is needed?*** The mathematical proof of the threshold theorem is founded on a number of assumptions treated as axioms:

1. Qubits can be prepared in the $|00000\ldots00\rangle$ state. New qubits can be prepared on demand in the state $|0\rangle$,
2. The noise in qubits, gates, and measurements is uncorrelated in space and time,
3. No undesired action of gates on other qubits,
4. No systematic errors in gates, measurements, and qubit preparation,
5. No undesired interaction between qubits,
6. No "leakage" errors,
7. Massive parallelism: gates and measurements are applied simultaneously to many qubits, and some others.

While the threshold theorem is a remarkable mathematical achievement, one would expect that these underlying assumptions, considered as axioms, would undergo a close scrutiny to verify that they can be reasonably approached in the physical world. Moreover, the term "reasonably approached" should have been clarified by indicating with what precision each assumption should be fulfilled. So far, this has never been done.

Note that in a technical instruction one does not say things like: "Make an Hadamard transformation on $N$ qubits", this does not make any sense for the engineer. Instead, one should say: "Rotate each of your million spins around the $y$-axis (defined with a precision $\varepsilon_1$) by 90 (plus or minus $\varepsilon_2$) degrees. Initially, the spin should be directed along $z$ with a precision $\varepsilon_3$. Also, take care that the disturbance of other spins by this action is less than $\varepsilon_4$, and that the spin-spin interaction in dimensionless units is less than $\varepsilon_5$."

The Future Quantum Engineer will need *this* kind of instruction together with the values of all the epsilons to understand whether the quantum computer is feasible.

***My own Axiom*** In the physical world, continuous quantities can be neither measured nor manipulated exactly. In the spirit of the purely mathematical language of the quantum computing theory, this can be formulated in the form of the following

**Axiom 1** No continuous quantity can have an exact value.

**Corollary** No continuous quantity can be exactly equal to zero.

To a mathematician, this might sound absurd. Nevertheless, this is an unquestionable reality of the physical world we live in. Note that *discrete* quantities, like the number of students in a classroom or the number of transistors in the on-state, *can* be known exactly and *this* is what makes the great difference between the digital computer and an analog computer, whether classical or quantum.

Axiom 1 is crucial whenever one deals with continuous variables. Thus if we devise some technical instruction, each step should contain an indication of the needed precision. Do not tell the engineer: "Make this angle 45°, and then my proposed vehicle will run as predicted, under the assumption that the road is flat".

This makes no sense! Tell him instead: "Make this angle $45° \pm 0.1°$, and then my proposed vehicle will run as predicted, provided the roughness of the road does not exceed 10 nm" (or 10 cm, whatever the theory says). Only then the engineer will be in a position to decide whether this is possible or not.

All of this is quite obvious, and nobody is going to believe that even in a thousand years somebody will manage to make the angle exactly $45°$ and provide an absolutely flat road to implement our invention.

***Precision of quantum amplitudes***  Apparently, things are not so obvious in the magic world of quantum mechanics. There is a widespread belief that the $|0\rangle$ and $|1\rangle$ states "in the computational basis" are something *absolute*, akin to the on/off states of an electrical switch, or of a transistor in a digital computer, but with the advantage that one can use quantum superpositions of these states, see Fig. 4.1 in Chap. 1. It is sufficient to ask: "with respect to which axis do we have a spin-up state?" to see that there is a serious problem with such a point of view.

It should be stressed once more that the coordinate system, and hence the computational basis, cannot be exactly defined, and this has nothing to do with quantum mechanics. Suppose that, again, we have chosen the $z$ axis towards the Polar Star, and we measure the $z$-projection of the spin with a Stern-Gerlach beam-splitter (a device using an inhomogeneous magnetic field to separate the spin-up and spin-down electrons). There will be inevitably some (unknown) error in the alignment of the magnetic field in our apparatus with the chosen direction. Thus, when we measure some quantum state and get (0), we never know exactly to what state the wave function has collapsed.

Presumably, it will collapse to the spin-down state with respect to the (not known exactly) direction of the magnetic field in our beam-splitter. However, with respect to the chosen $z$ axis (this direction is not known exactly either) the wave function will always have the form $a|0\rangle + b|1\rangle$, where hopefully $|b| \ll 1$. Another measurement with a similar instrument or a consecutive measurement with the same instrument will give a different value of $b$.

Quite obviously, the unwanted admixture of the $|1\rangle$ state is an error that *cannot be corrected*, since (contrary to the assumption 1 above) we can never have the standard *exact* $|0\rangle$ and $|1\rangle$ *states* to make the comparison. Thus, with respect to the consequences of imperfections, the situation is quite similar to what we have in classical physics. The classical statement "the exact direction of a vector is unknown" is translated into quantum language as "there is an unknown admixture of unwanted states".

The pure state $|0\rangle$ can never be achieved, just as a classical vector can never be made to point *exactly* in the $z$ direction, and for the same reasons–after all quantum measurements and manipulations are done with classical instruments.

Clearly, the same applies to *any* desired state. Thus, when we contemplate the "cat state" $(|0000000\rangle + |1111111\rangle)/\sqrt{2}$, we should not take the $\sqrt{2}$ too seriously, and we should understand that *some* (maybe small) admixture of all other 126 possible states of 7 qubits must be necessarily present.

Exact quantum states do not exist.

Some admixtures of all possible states to any desired state are unavoidable.

This fundamental fact described by Axiom 1 (nothing can be *exactly* zero!) should be taken into account in any prescriptions for quantum error correction. Note, that the "digitization" of noise, which is the cornerstone of the existing error-correcting schemes, is based on the *opposite* assumption, that the exact $|0\rangle$ and $|1\rangle$ states *can* be prepared.

In my opinion, it is premature to accept the threshold theorem as a proven result. The state of a quantum computer is described by the monstrous wave function with its (at least) $10^{300}$ amplitudes, all of which are continuously changing complex variables. If left alone, this wave function will completely deteriorate during $1/N$ of the relaxation time of an individual qubit, where $N \sim 10^3 - 10^6$ is the number of qubits within the computer.[4]

It is absolutely incredible, that by applying external fields, which cannot be calibrated perfectly, doing imperfect measurements, and using converging sequences of "fault-tolerant", but imperfect, gates from the universal set, one can continuously repair this wave function, protecting it from the random drift of its $10^{300}$ amplitudes, and moreover make these amplitudes change in a precise and regular manner needed for useful quantum computations. And all of this on a time scale greatly exceeding the typical relaxation time of a single qubit.

The existing prescriptions for fault-tolerant computation are rather vague, and the exact underlying assumptions are not always clear. It seems likely that the (theoretical) success of fault-tolerant computation is due not so much to the "quantum tricks", but rather to the backdoor introduction of ideal (flawless) elements in an extremely complicated construction. Previously, this view was expressed by Kak [50].

I reiterate my main points:

1. The hopes for scalable quantum computing are founded entirely on the "threshold theorem": once the error per qubit per gate is below a certain value, indefinitely long computations are possible.
2. The mathematical proof of the threshold theorem heavily relies on a number of assumptions supposed to be fulfilled *exactly*, as axioms.
3. In the physical world nothing can be exact when one deals with continuous quantities. For example, it is not possible to have *zero* interaction between qubits, it can only be *small*.

---

[4] Spin relaxation, i.e. the disappearance of the initial non-equilibrium spin polarization due to various natural causes was extensively studied for a very long time in gases, liquids, and solids. Depending on the system, the temperature, and the experimental conditions, the spin relaxation time $\tau_s$ for electrons usually varies from nanoseconds to milliseconds. As a rule, the decay of the *average* spin polarization is studied. However, it can be shown that the *general* nonequilibrium state of a system of $N$ spins decays $N$ *times faster*. Thus, if the average spin relaxation time is 1 s (which is enormously long!), but your system consists of $N = 1000$ spins, the initial spin state of the whole system will become unrecognizable after 1 ms. This gives us the upper time limit for performing our entire quantum algorithm.

4. Since the basic assumptions cannot be fulfilled *exactly*, the question is: What is the required *precision* with which each assumption should be fulfilled?
5. Until this crucial question is answered, the prospects of scalable quantum computing will remain very doubtful.

Dorit Aharonov, one of the authors of the threshold theorem, wrote in 1998 [51]: "In a sense, the question of noisy quantum computation is theoretically closed. But a question still ponders our minds: Are the assumptions on the noise correct?"

That's a very wise concern! Indeed, the question is closed *in the sense* that, based on a number of axioms, the theorem is proved. However, if by "noise" we mean *all* sort of uncertainties and undesired disturbances inevitably occurring in reality, then the assumptions on the noise are *not* correct, because *some* unavoidable noise is assumed to be absent, and *some* quantum states are assumed to be *exact*. In this *other sense*, the question remains wide open. Until somebody specifies the required precision with which various assumptions should be approached, the prospects of scalable quantum computing will remain very far from being clear.

A detailed description of the fault-tolerant computation rules, involving permanent verifications of the quantum states and repetitions of every our action, has been given by Preskill [45]. I don't find this description clear and/or convincing enough. Taking in account the continuous nature of random qubit rotation and gate inaccuracies, even with all the verifications and repetitions, there seems to be no way to avoid small admixtures of unwanted states to any desired state.

In fact, the pure spin-up state can never exist in reality (one reason is that we never know the exact direction of the $z$ axis). Similarly in the classical world we can never have a pointer looking *exactly* in the $z$ direction. Generally, no desired state can ever be achieved *exactly*, rather, whatever we do, we will *always* have an admixture of unwanted states, more or less rich. One can never have an exact $(|0\rangle + |1\rangle)/\sqrt{2}$ state, let alone more complicated "cat" states like $(|0000000\rangle + |1111111\rangle)/\sqrt{2}$. Such mathematical abstractions must be used with extreme caution when discussing the role of errors and inaccuracies.

When the small undetected and unknown admixture of unwanted states together with the "useful" state is fed into the subsequent stages of the quantum network, it is most likely that the error will grow exponentially in time. Accordingly, the crash time will depend only logarithmically on the initial error value.

This is what happens when one tries to reverse the evolution of a gas of hard balls in a box. At a given moment one reverses the direction of all the velocities, but oops, the gas never returns to its initial state, as theoretically it should (even in computer simulation, let alone reality). The reason is that however small the initial (and subsequent) computer errors are, they will increase exponentially with time (the Lyapunov exponent). It is a great illusion to think that things are different in quantum mechanics.

Related to this, there is another persistent misunderstanding of quantum mechanics, which plagues the quantum error correction literature. Using quite classical language, one says that the qubit "decoheres" with probability $p = \sin^2\theta$, instead of saying: the qubit is in the state $\psi = \cos\theta|0\rangle + \sin\theta|1\rangle$. It makes only a semantic difference if we are going to immediately *measure* the qubit, since the probability of finding it in a state $|1\rangle$ is indeed $p$. However, this language becomes wrong if we consider some further evolution of our qubit with a unitary matrix $R$. The common thinking (applied, for example, for estimating the noise threshold) is that we will have the state $R|0\rangle$ with probability $1-p$, and the state $R|1\rangle$ with probability $p$. In reality, we will have the state $R\psi$, and it is not the same thing. The former line of reasoning gives the probability of measuring $|0\rangle$ in the final state as $(1-p)|\langle 0|R|0\rangle|^2 + p|\langle 0|R|1\rangle|^2$, while the latter (and correct) one will give $|\langle 0|R|\psi\rangle|^2$, and these results are very different. As an exercise, the reader can take for $R$ a rotation of our qubit around the $x$ axis by some angle and compare the results. A quantum-mechanical surprise lies in store.

In quantum mechanics, one cannot calculate probabilities by considering what happens to ideal states. Instead, one must look at the evolution of the *real* states which always differ from ideal ones by some admixture of unwanted states.

Another point is that the finite time needed to do anything at all, is usually not taken into account. According to the proposed procedure of error correction [45], measuring the syndrome and obtaining (000) indicates the correct state that requires no further action. In fact, *while* we were making our measurements, the data qubits have experienced their random rotations. And no matter how many times we repeat the measurements this will happen again and again. Then why bother with error correction?

Alicki [2] has made a mathematical analysis of the consequences of finite gate duration. He writes: "…unfortunately, the success of existing error correction procedures is due to the discrete in time modeling of quantum evolution. In physical terms discrete models correspond to unphysical infinitely fast gates".

***Designing perpetual motion machine of the second kind***   This is certainly *not* equivalent to achieving fault-tolerant quantum computation, during which we will put some energy into the system by applying external fields and performing measurements.

However there is a certain similarity between the two problems in the sense that what we are trying to do is to maintain a reversible evolution of a large system with many degrees of freedom in the presence of noise and using noisy devices [52]. People, who have had the opportunity of considering projects of perpetual motion machines of the second kind (*not* violating the energy conservation law), know their basic principle: insert at least *one* ideal (i.e. not sensitive to thermal fluctuations) element somewhere deep within a complicated construction. Finding and identifying such an ideal element may be a daunting task.

Naively, one starts with proposing a valve that preferentially lets through only fast molecules. Next, one understands, that the valve itself is "noisy", so that it will not work as expected. However, if one adopts the noise model, in which the valve is faulty with probability $p$ but works perfectly with a probability $1-p$, or makes a sophisticated construction involving many valves connected by wheels and springs, and if just *one* of these elements is considered as *ideal* (or even working perfectly with some probability), one can immediately arrive at the conclusion that a perpetual motion machine feeding on thermal energy is possible.

This lesson should make us extremely vigilant to the explicit or implicit presence of ideal elements within the quantum error-correcting theoretical schemes.

It is absolutely incredible, that by applying external fields, which cannot be calibrated perfectly, doing imperfect measurements, and using converging sequences of "fault-tolerant", but imperfect, gates from the universal set, one can continuously repair this wave function, protecting it from the random drift of its $10^{300}$ amplitudes, and moreover make these amplitudes change in a precise and regular manner needed for quantum computations. And all of this on a time scale greatly exceeding the typical relaxation time of a single qubit.

# Chapter 5
# More Problems with Quantum Computing

*Energy* The notion of *energy* is of primordial importance in physics, both classical and quantum. However, quite astonishingly, it is not part of the rather restricted quantum computing theoretical vocabulary consisting of "qubits, gates, errors, and measurements". Apparently, for this reason, some very important points have been missed.

In specific proposals for quantum computing with spins in quantum dots, trapped ions, etc., the energies of quantum states are obviously introduced as an important issue. However, these proposals mostly concentrate on initialization, implementing gates, and read-out, and do not, to my knowledge, discuss the consequences of the free evolution of the quantum system that is addressed below.

***The undesired free evolution of the quantum computer*** In theory, so long as there is no noise and no gates are applied, an arbitrary superposition of $2^N$ states of $N$ qubits remains intact, implying that any such superposition is a *stationary state*. This is true only under the condition that the *energies of all the states* in our superposition are *exactly equal*. Otherwise, there will be a free evolution of the system defined by its energy spectrum [53, 54], see also Chap. 3. In turn, the energy spectrum is determined by the external fields and by the interaction between qubits.

If the energy difference between the states $|1\rangle$ and $|0\rangle$ of a single qubit is $\Delta E = \hbar\Omega$, like for a spin in magnetic field (see Chap. 3), and at $t = 0$ we create the state $(|0\rangle + |1\rangle)/\sqrt{2}$, then at later times the state will evolve as $(|0\rangle + \exp(-i\Omega t)|1\rangle)/\sqrt{2}$, describing the spin precession in the plane perpendicular to magnetic field with frequency $\Omega$. In the general case of $N$ qubits, the dynamic of the system will be characterized by many harmonics of $\Omega$.

In fact, things are still worse, because, the external magnetic field cannot be absolutely homogeneous and also there necessarily will be some interaction between our qubits. Thus the values of $\Omega$ for different qubits cannot possibly be *exactly* equal (in spectroscopy, this is described as *inhomogeneous broadening*), so that the spectrum of our system will not simply consist of all harmonics of $\Omega$ from 1 to

M. I. Dyakonov, *Will We Ever Have a Quantum Computer?*,
SpringerBriefs in Physics, https://doi.org/10.1007/978-3-030-42019-2_5

$N-1$. Rather, it will have a quasi-continuous character resulting in a (quasi-) chaotic behavior of the system's wave function $\Psi(t)$.

Thus, it is extremely naive to think that our quantum computer is going to sit still waiting for us to apply our "gates", or perform measurements. The undesired free evolution consisting in a free rotation of all qubits with a wide frequency spectrum can hardly be described as noise, and it is not obvious at all how the Future Quantum Engineer is supposed to deal with it. It is quite amazing that the *energies* of $|0\rangle$ and $|1\rangle$ states and the elementary quantum-mechanical property described above are never a subject of discussion in the abstract theory of quantum computing.

Consider, for example, the task of putting a thousand of spins at different locations, all initially aligned in the $|0\rangle$ states by an external magnetic field, in the final state $\psi = (|1\rangle + |0\rangle)/\sqrt{2}$, by applying $\pi/2$ microwave pulse. Supposing that we have succeeded, consider the requirements for phase matching needed to perform at a later time the inverse operation of returning all spins back to the state $|0\rangle$.

***The physical quantum computer as a non-linear system*** [53, 54]. The *abstract* universal quantum computer, as first defined by Deutsch [14], is an assembly of qubits, with which certain well defined operations are feasible. However, since these operations will not fall from the sky, the eventual *physical* quantum computer will be an assembly of qubits PLUS a monstrously complex and sophisticated classical apparatus, needed to efficiently control many thousands, or maybe even millions, of qubits. To understand this, it is sufficient to visit a lab and take a look at the experimental setup used for controlling 3 qubits.

Now, Quantum Mechanics is a linear theory and when a unitary gate $U$ acts on the state $a|0\rangle + b|1\rangle$, one gets $aU|0\rangle + bU|1\rangle$. However *measurements*, an indispensible part of the quantum computing procedure, are *not* linear operations. Measurement-based error correction transforms the state $a|0\rangle + b|1\rangle$ to $|0\rangle$ or $|1\rangle$, depending on what term is considered to represent an error, and this is not a linear operation. The required equipment by itself is not a linear device either. Thus the whole machine is a huge and strongly nonlinear construction, which generally will exhibit instabilities and chaotic behavior.

Taking into account the enormous number of control parameters within a quantum computer, it is absolutely not clear whether stable large-scale quantum computing is achievable, and if yes, under what conditions.

This is an important, but totally unexplored issue. Meanwhile there exists a vast field of applied mathematics dealing with instabilities and control of nonlinear systems, see for example Refs. [55, 56]. This literature might be of some interest to QC theorists.

***Experimental studies*** Related to the idea of quantum computing make only a tiny part of the whole QC literature. They represent the *nec plus ultra* of the modern experimental technique, they are extremely difficult and inspire respect and admiration.

The goal of such proof-of-principle experiments is to show the possibility of manipulating small numbers of two-level quantum systems serving as qubits, to realize the basic quantum operations, as well as to demonstrate elements of the Shor's factoring algorithm and error correction by encoding.

**Factoring 15**   The first experiment reporting factoring 15 by Shor was reported by Vandersypen et al. [57] using the liquid nuclear magnetic resonance (NMR) technique. All the gates were implemented by microwave pulses applied within about 1 s, which is less than the nuclear decoherence time. The obtained NMR spectra corresponded very well to the predictions of Shor's procedure.

Lanyon et al. [58] performed the same task in an optical experiment using the polarization of 4 photons, while Lucero et al. [59] used Josephson qubits: "...we run a three-qubit compiled version of Shor's algorithm to factor the number 15, and successfully find the prime factors 48% of the time".

BUT: in these experiments the so-called *compiled* version of the Shor's algorithm was used. As it was shown by Beckman et al. [60], the full algorithm can factor a $k$-bit number using $72\,k^3$ elementary quantum gates; e.g., factoring 15 requires 4608 gates operating on 21 qubits. Recognizing that those requirements are *well beyond* the existing experimental possibilities, Beckman et al. introduced a *compiling* technique which exploits properties of the number to be factored, allowing exploration of Shor's algorithm with a vastly reduced number of resources.

One might say that this is a sort of (innocent?) cheating: knowing in advance that $15 = 3 \times 5$, we can take some shortcuts, which would not be possible if the result were not known beforehand.

All the existing experimental testing of Shor's algorithm use this simplified approach. In a very remarkable work of Martin-López et al. [61] the same approach allowed for the first time to factor 21 in an optical experiment, where an iterative procedure of recycling a single qubit was successfully implemented.

*Honest factoring 15* by *using the full Shor's algorithm is still well beyond the reach of experimental possibilities* (as I have predicted in 2001 [53]). Thus it seems that for the foreseeable future we need not worry about the security of cryptography codes based on the difficulty in factoring very large numbers, which are products of hundred-digit primes.

**Error correction**   While there is a huge amount of theoretical work devoted to quantum error correction, it is safe to say that up to now the experimental error correction is non-existent. This is why, the initial idea of *quantum annealing* put forward by the D-wave Systems Company becomes more and more appealing: instead of fighting noise, we *use* it, since precisely because of noise and energy dissipation, the quantum system tends to its ground state. This is not the universal quantum computer, everyone was talking about during all these years, but nevertheless such studies might give some interesting and potentially useful results.

**An optimistic look into the future**   Since the required precision of quantum gates, measurements, etc. still has not been established, let us assume a highly optimistic scenario that in 10, or 20–30, or 1000 years, the technology will advance to such a perfection that all technical and other problems are resolved and we are finally able to realize the dream of factoring numbers $\sim 10^{100}$ by Shor's algorithm. The Future Quantum Engineer will then ask for a *technical instruction*: what should be done, in what order, and with what precision. At present, it is not clear at all, how and when such an instruction can appear.

If the instruction for factoring one-hundred-digit numbers is too difficult to develop, maybe somebody could produce a technical manual for factoring 15 with error correction? (Full Shor's algorithm, please, no "compiled" versions).

Assuming that it will be created some day, it should be noted that the Future Quantum Engineer will not be able to check the state and adjust the functioning of his quantum computer because this requires knowing the values of (at least) $10^{300}$ quantum amplitudes, which is impossible.

So, he will have to just follow his instructions, construct the hardware, introduce the number to be factored as input, run the machine, and hope for the best: that the final measurement of 1000 logical qubits will give the correct answer.

What if it does not? How is the Future Quantum Engineer supposed to find the bug(s)?

A similar problem was put forward by Levin [9]:

> QC proponents often say they win either way, by making a working QC or by finding a correction to Quantum Mechanics. ... Consider, however, this scenario. With few $q$-bits, QC is eventually made to work. The progress stops, though, long before QC factoring starts competing with pencils. The QC people then demand some noble prize for the correction to the Quantum Mechanics. But the committee wants more specifics than simply a nonworking machine, so something like observing the state of the QC is needed. Then they find the Universe too small for observing individual states of the needed dimensions and accuracy. (Raising sufficient funds to compete with pencil factoring may justify a Nobel Prize in Economics.) ... So, what thought experiments can probe the QC to be in the state described with the accuracy needed? I would allow to use the resources of the entire Universe, but not more!

*Challenge*  In the case that even the simplest task of factoring 15 does not arouse any interest in the QC community, what about quantum computing the identity $1 = 1$, meaning that we just want to use the methods of error correction to store a given state of $N$ qubits for a time exceeding the decoherence time, say, by a factor of 100 or 1000 (quantum memory).

Let us start by $N = 1$, a challenge that I posed in 2006 [62]. Storing just one qubit in the presence of noise and gate inaccuracies is obviously the simplest meaningful problem in the field of fault-tolerant quantum computing. Remember, we are *not* talking about an experimental demonstration, which might be achieved not so soon, if ever. It is a purely theoretical question, and its interest lies in the fact that since the number of qubits involved is not very large, it might be possible to simulate the process on a laptop.

Presumably, to maintain our single qubit close to its initial state $a|0\rangle + b|1\rangle$ with given $a$ and $b$, a certain sequence of operations (with possible branching depending on the result of intermediate measurements) should be applied periodically.

*Provide a full list of these elementary operations*, so that anybody can use his PC to check whether qubit storage really works in the presence of noise and under the condition that both gates and measurements are imperfect, and what degrees of imperfections and noise are acceptable. Presumably, to maintain our single qubit close to its initial state, a certain sequence of gates (with possible branching depending on the result of intermediate measurements) should be applied periodically. The Future Quantum Engineer will certainly need such a list!

*If* it works, this demonstration would be a convincing, though partial, proof that the idea of fault-tolerant quantum computation is sound. So far, this challenge has never been met.[1]

The task of maintaining a single qubit in memory is much simpler and, once the list of required operations is provided, it hopefully can be simulated in a straightforward manner by using a normal computer. I predict that the result will be discouraging.

---

[1]The aversion of QC theorists to demonstrating how their methods are supposed to work in the simplest situations is reminiscent of the magician episode described by Twain [63]. The magician could easily tell what the Emperor of the East was doing, but could not guess what the Yankee was doing with his right hand. If you don't know how to protect from errors just one qubit, how can you talk about scalable fault-tolerant quantum computation with millions of qubits?

# Chapter 6
# Related Issues and Conclusions

*Mathematics and physical reality* Besides having an enormous intrinsic value, mathematics is indispensable for understanding the physical world, as well as for all practical human activities. However, there are many ways of abusing mathematics. For example, you want to discuss the very complex phenomenon of *love*. If you are a poet or a sexologist or, better still, if you have some experience of your own, you may have chances to say something reasonable. But if the only thing you have up your sleeve is to write the Hamiltonian of the couple as $H = H_1 + H_2 + V$, with $V$ being a sum of products of operators belonging to the subspaces 1 and 2, and then, *under certain assumptions*, you prove some theorems, quite obviously your rigorous results will be both wrong and irrelevant.

And the reason is that you simply have no idea even about $H_1$, and still less about $H_2$, not to mention $V$. As a consequence, whatever your assumptions are, they are groundless [64].

Another example, nearer to our subject, can be found in Jaroslav Hašek's masterpiece [65]. The good soldier Švejk spends some time in a madhouse, where one of the professors among the patients tries to convince everybody that "Inside the terrestrial globe there is another globe of a much greater diameter".

In fact, that mad professor's claim is consistent with the well-known mathematical fact that, in certain metric spaces, a ball of a bigger radius may be properly contained in one of a smaller radius. Another (perfectly rigorous) result, which he also anticipated in a way, is the famous Banach-Tarski paradox that was discovered 10 years later [66]; see the very clear article in Wikipedia [67].[1]

A version of the Banach-Tarski theorem states that, given a small ball and a huge ball in the usual 3D Euclidean space, either one can be partitioned into pieces and then reassembled into the other. Note that the number of pieces is *finite*, and the reassembly process consists in moving them around, using only translations and rotations (but no stretching). Thus, as Wikipedia puts it, "a pea can be chopped up and reassembled into the Sun".

---

[1]I thank Konstantin Dyakonov for bringing these results to my attention.

© The Author(s), under exclusive license to Springer Nature Switzerland AG 2020
M. I. Dyakonov, *Will We Ever Have a Quantum Computer?*,
SpringerBriefs in Physics, https://doi.org/10.1007/978-3-030-42019-2_6

The interested reader must learn about metric spaces, non-measurable sets, and other mathematical technicalities in order to understand how it is possible that a rigorously proved theorem contradicts our common sense, and whether we should revise our common sense accordingly. (The short answer is *no*, because our common sense is based on the structure of the surrounding physical world, while the axioms behind the theorem are not.)

Now, imagine some society on another planet, where an army of scientific journalists not familiar with the technicalities cites the distinguished experts, affirming (quite correctly, *in some sense*) that *it has been proved* that one can build a full-scale skyscraper on the basis of its 1000:1 model without using any additional material, and more importantly, that the same applies to a tank or a submarine. Scientists in top-secret labs are developing the technical procedures, and many promising materials have been already proposed, such as endohedral fullerenes, ionic Wigner crystals, and some others.

That's what is happening in some places of our planet with respect to the threshold theorem: "The theory of fault-tolerant quantum computation establishes that a noisy quantum computer can simulate an ideal quantum computer accurately. In particular, the quantum accuracy threshold theorem asserts that an arbitrarily long quantum computation can be executed reliably, provided that the noise afflicting the computer's hardware is weaker than a certain critical value, the *accuracy threshold*" [23]. In other words, the possibility of scalable quantum computing has been rigorously proved. As we have seen, this is not true.

No theorem can be proved without a set of axioms on which it relies. It is not the mathematician's concern whether his axioms describe the physical reality correctly or not. However, this should be the *main* concern for those who want to apply the theorem to the physical world. This question lies outside mathematics and the only way to solve it is to consult experiment and the "engineers", who are absolutely not familiar with and will never believe in "decoherence-free subspaces" and "approaching with arbitrary precision any unitary transformation on $N$ qubits by an appropriate number of gates from the universal set".

***Quantum computing as a sociological problem***   In fact, quantum computing is not so much a scientific, as a sociological problem which has expanded out of all proportion due to the US system of funding scientific research (which is now being copied all over the world). While having some positive sides, this system is unstable against spontaneous formation of bubbles and mafia-like structures. It pushes the average researcher to wild exaggerations on the border of fraud and sometimes beyond. Also, it is much easier to understand the workings of the funding system, than the workings of Nature, and these two skills only rarely come together.

The QC story says a lot about human nature, the scientific community, and the society as a whole, so it deserves profound psycho-sociological studies, which should begin right now, while the main actors are still alive and can be questioned.

A somewhat similar story can be traced back to the 13th century when Hodja [68] made a proposal to teach his donkey to read and obtained a 10-year grant from the local Sultan. For his first report he put breadcrumbs between the pages of a big book, and demonstrated the donkey turning the pages with his hoofs. Although no scientific publications followed, this was a promising first step in the right direction.

Nasreddin was a wise but simple man, so when asked by friends how he hopes to achieve his goal, he answered: "My dear fellows, before ten years are up, either I will die or the Sultan will die. Or else, the donkey will die."

Had he the modern degree of sophistication, he could say, first, that there is no theorem forbidding donkeys to read. And, since this does not contradict any known fundamental principles, the failure to achieve this goal would reveal new laws of Nature. So, it is a win-win strategy: either the donkey learns to read, or new laws will be discovered.[2]

Second, he could say that his research may, with some modifications, be generalized to other animals, like goats and sheep, as well as to insects, like ants, gnats, and flies, and this will have a tremendous potential for improving national security: these beasts could easily cross the enemy lines, read the secret plans, and report them back to us.

The modern version of these ideas is this love-song for military sponsors: "The transistors in our classical computers are becoming smaller and smaller, approaching the atomic scale. The functioning of future devices will be governed by quantum laws. However, quantum behavior cannot be efficiently simulated by digital computers. Hence, the enormous power of quantum computers will help us to design the future quantum technology."

This may look good to a project manager, or in some science digest magazine, but for anyone who understands something about simulation, quantum laws, transistors, and atoms, this does not make any sense at all.

I believe that, in spite of appearances, the quantum computing story is nearing its end, not because somebody provides a mathematical proof that it is impossible, but rather because 25 years is a typical lifetime of any big bubble in science, because too many unfounded promises have been made, because people get tired and annoyed by almost daily announcements of new "breakthroughs" [70], because all the tenure positions in quantum computing are already occupied, and because the proponents are growing older and less zealous, while the younger generation seeks for something new.

The ground-breaking results of Benioff [11], Deutsch [14], Shor [15], and some others, will certainly remain for a long time because new and audacious ideas are always valuable, whether they lead to practical results, or not (obviously, this does not apply to the major part of the huge quantum computing literature). However it should be fully realized that the practical implementation of these ideas requires complete control over all the intimate details of a many-particle quantum system characterized by a huge number of continuous parameters.

---

[2]This kind of reasoning is quite common in our days: "Because there are no known fundamental obstacles to such scalability, it has been suggested that failure to achieve it would reveal new physics" [69].

There is an intermediate link between a quantum and a digital computer, which is a classical analog machine, like a system of $10^5$ oscillators or classical magnetic moments. Why not try to build such a machine first, and then, when we see that it works, try to accomplish the next, orders-of-magnitude more difficult task of constructing a quantum computer with $10^5$ quantum spins? It is very hard to believe that even such a classical analog machine of sufficient complexity will ever work.

*Conclusions* The gigantic field of "quantum computing" with many thousands of active researchers, hundreds of thousands of publications, numerous conferences and workshops, daily announcements of new breakthroughs, and many, many billions of dollars spent, has been triggered by Shor's invention of his famous algorithm for factoring extremely large numbers (thus eventually opening the door for quantum computers to break security codes), and by the developments of methods for quantum error correction, which is generally considered as being *absolutely indispensable*.

The worldwide quantum computing euphoria and the general excitement are going on already for a quarter of a century! Before engaging further for another 25 years, it might be wise to have a look at the achievements reached to date during this period.
The observable outcome can be summed up as follows:

- Factoring the number 15 by Shor's algorithm is still not possible.
- Error correction has still never been achieved, even on a very small scale.
- No quantum device exists, capable of doing elementary arithmetic, like $3 \times 5$, or $3 + 5$.

Thus, after a quarter of a century, there are absolutely NO meaningful results in quantum computing!

- The only working quantum machines to date are those introduced by the D-wave Systems company in 1999, and currently intensely studied and developed by Amazon, Google, IBM, and other tech giants, as well as by the D-wave company itself. These machines can perform *quantum annealing* but so far are not capable of error correction and thus are NOT quantum computers in the original sense of this term. They too are not capable of factoring 15, nor calculating $3 + 5$. However, they are interesting from the scientific point of view and allow to obtain some valuable results [30–33].
- With no clear reasons to believe that this situation is going to change during the next 25 years, the perspectives of quantum computing appear to be extremely doubtful.

Skepticism is a normal and healthy attitude in science, as opposed to religion, and it is for the believer to give a convincing proof that the anticipated miracle is about to happen. After 25 years of considerable efforts by many thousands of researchers, such a proof is still absent.
The unprecedented level of hype and of unfounded promises accompanying the QC enterprise is not a good sign either, as well as the multitude of mostly quite irresponsible proposals of the type "quantum computing with…". It is a real pity that they never contain, as they should, the famous Landauer's disclaimer [7]:

This scheme, like all other schemes for quantum computation, relies on speculative technology, does not in its current form take into account all possible sources of noise, unreliability and manufacturing error, and probably will not work.

In riding a bike, after some training, we learn to successfully control 3 degrees of freedom: the velocity, the direction, and the angle that our body makes with respect to the pavement. A circus artist manages to ride a one-wheel bike with 4 degrees of freedom. Now, imagine a bike having 1000 (or rather $2^{1000}$) joints that allow free rotations of their parts with respect to each other. Will anybody be capable of riding this machine?

It appears pretty obvious that the answer to the question in title is:

*No, we will never have a quantum computer.* Instead, we might have some special-task (and outrageously expensive) quantum devices operating at millikelvin temperatures.

The saga of quantum computing is waiting for a profound sociological analysis, and some lessons for the future should be learnt from this fascinating adventure.

This book is based on my previous papers [53, 54, 62, 71–74].

**Acknowledgements**  It is my great pleasure to thank Serge Luryi, Jimmy Xu, and Alex Zaslavsky, the organizers of the "Future Trends in Microelectronics" conference series, for inviting me to give talks about quantum computing and for sharing my views. I also thank my family for help and support, especially my son Mikhail for drawing the picture at the frontispiece.

# References

1. https://www.forbes.com/sites/ibm/2020/01/16/the-quantum-computing-era-is-here-why-it-mattersand-how-it-may-change-our-world/#45629785c2b1
2. Alicki, R.: Quantum error correction fails for Hamiltonian models. Fluct. Noise Lett. (2006). arxiv.org/quant-ph/0411008
3. Haroche, S., Raymond, J.-M.: Quantum computing: dream or nightmare. Phys. Today **49**, 8, 51 (1996)
4. 't Hooft, G.: Quantum gravity as a dissipative deterministic system. Class. Quantum Grav. **16**, 3263 (1999). gr-qc/9903084.
5. Kak, S.: General qubit errors cannot be corrected. Inf. Sci. **152**, 195 (2003)
6. Kalai, G.: How quantum computers fail: quantum codes, correlations in physical systems, and noise accumulation. arXiv:1106.0485. https://www.quantamagazine.org/gil-kalai-argument-against-quantum-computers-20180207/
7. Landauer, R.: The physical nature of information. Phys. Lett. A **217**, 188 (1996)
8. Laughlin, R.: https://www.youtube.com/watch?v=iYQSbV_BlI8
9. Levin, L.: Polynomial time and extravagant models (2003). The tale of one-way functions. Probl. Inf. Transm. **39**(1) (2003). https://arxiv.org/pdf/cs/0012023.pdf
10. Wolfram, S.: Undecidability and intractability in theoretical physics. Phys. Rev. Lett. **54**, 735 (1985)
11. Benioff, P.: The computer as a physical system: a microscopic quantum mechanical Hamiltonian model of computers as represented by Turing machines. J. Stat. Phys. **22**, 563 (1980)
12. Manin, Y.: Vychislimoe i nevychislimoe (Computable and Noncomputable). Sov. Radio, pp. 13–15 (1980) (in Russian)
13. Feynman, R.P.: Simulating physics with computers. Int. J. Theor. Phys. **21**, 467 (1982)
14. Deutsch, D.: Quantum theory, the Church-Turing principle and the universal quantum computer. Proc. Roy. Soc. Lond. A **400**, 97 (1985)
15. Shor, P.W.: Algorithms for quantum computation: discrete logarithms and factoring. In: Proceedings of the 35th Annual Symposium on Foundations of Computer Science, p. 124. IEEE Computer Society Press, Los Alamitos, CA (1994)
16. Grover, L.K.: A fast quantum mechanical algorithm for database search. In: Proceedings of the 28-th Annual ACM Symposium on the Theory of Computing, p. 212 (1996)
17. Shor, P.W.: Scheme for reducing decoherence in quantum computer memory. Phys. Rev. A **52**, 2493 (1995)
18. Steane, A.M.: Error correcting codes in quantum theory. Phys. Rev. Lett. **77**, 793 (1996)
19. Ofek, N., Petrenko, A., et al.: Extending the lifetime of a quantum bit with error correction superconducting circuits. Nature **536**, 441 (2016)
20. Calderbank, A.R., Shor, P.W.: Good quantum error-correcting codes exist. Phys. Rev. A **54**, 1098 (1996)

M. I. Dyakonov, *Will We Ever Have a Quantum Computer?*,
SpringerBriefs in Physics, https://doi.org/10.1007/978-3-030-42019-2

21. Aharonov, D., Ben-Or, M.: Fault-tolerant quantum computation with constant error rate. In: Proceedings of the 29th Annual ACM Symposium on the Theory of Computation, p. 176, ACM Press, New York (1998). arXiv:quant-ph/9611025, http://arXiv:quant-ph/9906129
22. Knill, E.: Extracting information from qubits-environment correlations. Nature **463**, 441 (2010)
23. Aliferis, P., Gottesman, D., and Preskill, J.: Accuracy threshold for post selected quantum computation. Quantum Inf. Comput. **8**, 181 (2008). arXiv:quant-ph/0703264
24. D-wave systems. https://en.wikipedia.org/wiki/D-Wave_Systems
25. https://docs.dwavesys.com/docs/latest/c_gs_2.html,  https://en.wikipedia.org/wiki/D-Wave_Systems#History
26. https://en.wikipedia.org/wiki/Pi_Josephson_junction
27. https://en.wikipedia.org/wiki/Josephson_effect
28. Likharev, K.K.: Dynamics of Josephson Junctions and Circuits. Gordon and Breach Publications, New York (1986)
29. King, A.D., Carrasquilla, J., et al.: Observation of topological phenomena in a programmable lattice of 1,800 qubits. Nature **560**, 456 (2018)
30. Chen, Y., Neill. C., et al.: Phys. Rev. Lett. **113**, 220502 (2014). https://arxiv.org/pdf/1402.7367.pdf
31. Arute, F., et al.: Quantum supremacy using a programmable superconducting processor. Nature **574**, 505 (2019). https://www.nature.com/articles/s41586-019-1666-5
32. https://www.quantamagazine.org/google-and-ibm-clash-over-quantum-supremacy-claim-20191023/
33. Foxen, B., et al.: Demonstrating a continuous set of two-qubit gates for near-term quantum algorithms. https://arxiv.org/abs/2001.08343
34. Quantum Manifesto. https://qt.eu/app/uploads/2018/04/93056_Quantum Manifesto_WEB.pdf
35. A quantum information science and technology roadmap, Part 1: quantum computation, report of the quantum information science and technology experts panel. http://qist.lanl.gov/qcomp-map.shtml
36. National academies of sciences, engineering, and medicine, quantum computing: progress and prospects. National Academies Press, Washington, DC (2018). https://doi.org/10.17226/25196
37. Bravyi, S.: Universal quantum computation with the $\nu = 5/2$ fractional quantum Hall state. Phys. Rev. A **73**, 042313 (2006). arXiv: quant-ph/0511178
38. Wilczek, F. (ed.): Fractional Statistics and Anyon Superconductivity. World Scientific, Singapore (1990)
39. Kitaev, A.Y.: Fault-tolerant quantum computation by anyons. Ann. Phys. **303**, 2 (2003). arXiv: quant-ph/9707021
40. Moore, J.E.: The birth of topological insulators. Nature **464**, 194 (2010)
41. Wilczek, F.: Majorana returns. Nat. Phys. **5**, 614 (2009)
42. Lian, B., Sun, X.-Q., et al.: Topological quantum computation based on chiral Majorana fermions. Proc. Nat. Acad. Sci. USA **115**(43), 10938 (2018)
43. Pade, J.: Quantum Mechanics for Pedestrians 1: Fundamentals. Springer (2014)
44. Shor, P.: Fault-tolerant quantum computation. In: 37th Symposium on Foundations of Computing, pp. 56–65. IEEE Computer Society Press (1996). arxiv.org/quant-ph/9605011
45. Preskill, J.: Fault-tolerant quantum computation. In: Lo, H.-K., Papesku, S., Spiller, T. (eds.) Introduction to Quantum Computation and Information, pp. 213–269. World Scientific, Singapore (1998). arxiv.org/quant-ph/9712048
46. Gottesman, D.: An introduction to quantum error correction. In: Lomonaco Jr., S.J. (ed.) Quantum Computation: A Grand Mathematical Challenge for the Twenty-First Century and the Millennium, pp. 221–235. American Mathematical Society, Providence, Rhode Island (2002). arxiv.org/quant-ph/0004072
47. Steane, A.M.: Quantum computing and error correction. In: Gonis, A., Turchi, P. (eds.) Decoherence and its Implications in Quantum Computation and Information Transfer, pp. 284–298. IOS Press, Amsterdam (2001). arxiv.org/quant-ph/0304016
48. Steane, A.M.: Overhead and noise threshold of fault-tolerant quantum error correction. Phys. Rev. A **68**, 042322 (2003)

49. Kribs, D., Laflamme, R., Poulin, D.: A unified and generalized approach to quantum error correction. Phys. Rev. Lett. **94**, 180501 (2005)
50. Kak, S.: General qubit errors cannot be corrected. Inf. Sci. **152**, 195 (2003). arxiv.org/quant-ph/0206144
51. Aharonov, D.: Quantum computation. In: Stauffer, D. (ed.) Annual Reviews of Computational Physics, p. 259. World Scientific, VI (1999). arXiv:quantph/98120
52. Because of this similarity, it is quite probable that the design and the theory of such machines will become the next hot topic, especially if a good name (beginning with the magic word "quantum") for this activity will be invented
53. Dyakonov, M.I.: Quantum computing: a view from the enemy camp. In: Luryi, S., Xu, J., Zaslavsky, A. (eds.) Future Trends in Microelectronics. The Nano Millennium, p. 307, Wiley (2002). arXiv:cond-mat/0110326
54. Dyakonov, M.I.: Prospects for quantum computing: extremely doubtful. Int. J. Mod. Phys.Conf. Ser. **33**, 1460357 (2014). arxiv:1401.3629
55. Sastry, S.: Nonlinear Systems: Analysis, Stability and Control. Springer (1999)
56. Gutzwiller, M.C.: Chaos in Classical and Quantum Mechanics. Springer, New York (1990)
57. Vandersypen, L.M.K., et al.: Experimental realization of Shor's quantum factoring algorithm using nuclear magnetic resonance. Nature **414**, 883 (2001)
58. Lanyon, B.P., et al.: Experimental demonstration of a compiled version of Shor's algorithm with quantum entanglement. Phys. Rev. Lett. **99**, 250505 (2007)
59. Lucero, E., et al.: Computing prime factors with a Josephson phase qubit quantum processor. Nat. Phys. **8**, 719 (2012). arXiv:1202.5707
60. Beckman, D., et al.: Efficient networks for quantum factoring. Phys. Rev. A **54**, 1034 (1996). arXiv:quant-ph/9602016
61. Martin-López, E., et al.: Experimental realization of Shor's quantum factoring algorithm using qubit recycling. Nat. Photonics, **6**, 773 (2012). arXiv:1111.4147
62. Dyakonov, M.I.: Is fault-tolerant quantum computation really possible? In: Luryi, S., Xu, J., Zaslavsky, A. (eds.) Future Trends in Microelectronics. Up the Nano Creek, p. 4. Wiley (2007). arXiv:quant-ph/0610117
63. Twain, M.: A Connecticut Yankee in King Arthur's Court, Chapter 24 (1889). http://www.gutenberg.org/files/86/86-h/86-h.htm
64. To put it bluntly, one should refrain from proving theorems about systems and phenomena of which one has no profound understanding
65. Hašek, J.: The Fateful Adventures of the Good Soldier Švejk During the World War (translated by Sadlon, Z.K.). SAMIZDAT (2007). amazon.com/Fateful-Adventures-Soldier-Svejk-During/dp/1585004286
66. Banach, S., Tarski, A.: Sur la décomposition des ensembles de points en parties respectivement congruents. Fundam. Math. **6**, 244 (1924)
67. www.wikipedia.org/wiki/Banach_Tarski
68. Nasreddin Hodja was a populist philosopher and a wise man believed to have lived around 13-th century during the Seljuq dynasty and remembered for his funny stories and anecdotes. The International Nasreddin Hodja fest is celebrated annually in July in Aksehir, Turkey every year
69. Knill, E.: Quantum computing. Nat. Phys. **463**, 441 (2010)
70. When a spin relaxation time of 10 ns is measured, instead of 1 ns previously, this is heralded as another breakthrough on the way to quantum computing
71. Dyakonov, M.I.: State of the art and prospects for quantum computing. In: Luryi, S., Xu, J., Zaslavsky, A. (eds) Future Trends in Microelectronics. Frontiers and Innovations, p. 266. Wiley (2013). arXiv:1212.3562
72. Dyakonov, M.I.: Revisiting the hopes for scalable quantum computation. JETP Lett. **98**, 514 (2013). arXiv:1210.1782
73. Dyakonov, M.: The case against quantum computing. IEEE Spectr. **56**(3), 24 (2019)
74. Dyakonov, M.I.: When will we have a quantum computer? Solid State Electron. **155**, 4 (2019)

# Index

© The Author(s), under exclusive license to Springer Nature Switzerland AG 2020          49
M. I. Dyakonov, *Will We Ever Have a Quantum Computer?*,
SpringerBriefs in Physics, https://doi.org/10.1007/978-3-030-42019-2

Printed in the United States
By Bookmasters